ATLAS
OF
VISUALIZATION
II

Editors-in-Chief
Yasuki Nakayama
Yoshimichi Tanida

Edited by
The Visualization Society of Japan

CRC Press
Boca Raton   New York   London   Tokyo

Acquiring Editor: Joel Claypool
Contact Editor: Ben Kato
Associate Editor: Felicia Shapiro
Project Editor: Jennifer Richardson
Marketing Manager: Susie Carlisle
Direct Mail Marketing Manager: Becky McEldowney
Prepress: Kevin Luong
Cover Designer: Shayna Murray
Manufacturing: Sheri Schwartz

Library of Congress Cataloging-in-Publication Data

Atlas of visualization, II / edited by The Visualization Society of
    Japan ; editors-in-chief, Yasuki Nakayama and Yoshimichi Tanida.
        p.   cm.
    Includes bibliographical references and index.
    ISBN 0-8493-2656-7 (alk. paper)
    1. Flow visualization. I. Nakayama, Y. (Yasuki), 1916-   .
    II. Tanida, Y. (Yoshimichi), 1931-   . III. Visualization Society
    of Japan.
    TA357.A876   1996
    681'.2—dc20

                                                                    96-13917
                                                                         CIP

Preface

Visualization is a new science where invisible information is made positively visible using visualization techniques and computers. Through such visible information new information is obtained which helps to clarify the phenomena.

To promote this science, the Visualization Society of Japan (VSJ) was established as an official society registered in the Government of Japan in 1990. This society is the developed version of the Flow Visualization Symposium which started in 1973.

The Atlas of Visualization was first published as the English Journal of the VSJ in 1992. In view of our experience preparing our first issue, the Editorial Committee was reorganized so that major countries are cordially solicited to have Editors in order to facilitate the committee management.

Now the VSJ has decided to publish the Atlas of Visualization Volume 2 with the same policy as the first one. This series will be published annually for the time being, but hopefully will increase its publishing frequency in the future.

The papers contained will be requested papers, recommended papers, and contributed papers. The contents are roughly divided into experimental visualization and computer-aided visualization. Their subjects will range from fluid flow to heat and mass transfer, acoustic energy, electromagnetism, chemical change, etc. Their fields will cover not only engineering and physics but also such disciplines as medical science, agriculture, oceanography, meteorology, and sports science.

Moreover, the Atlas of Visualization will be issued in full color so that complex phenomena can be understood more clearly.

Lastly, on behalf of the Editorial Committee, we would like to give our sincere acknowledgment to the authors who presented their valuable papers and photographs to this series for exchanging new knowledge through it.

Yasuki Nakayama
Yoshimichi Tanida
Editors-in-Chief

Y. Nakayama

Y. Tanida

Editorial Board

Contents

Frontispiece Illustrations

Figure 1   H. Yamashita, G. Kushida, and T. Takeno
   Nagoya University, Japan

Figure 2   J. D. Dale
   University of Alberta, Canada

Figure 3   G. H. Schnerr
   University of Karlsruhe, Germany

Figure 4   K. Takeishi
   Mitsubishi Heavy Industries, Ltd., Japan

Figure 5   T. Suzuki, Y. Aoyagi, H. Yokota
   Hino Motors, Ltd., Japan

Figure 6   Y. Ye
   China Aerodynamics Research and Development Center, China

Figure 7   K. Fujii
   Institute of Space and Aeronautical Science, Japan

Figure 8   N. Kasagi
   University of Tokyo, Japan

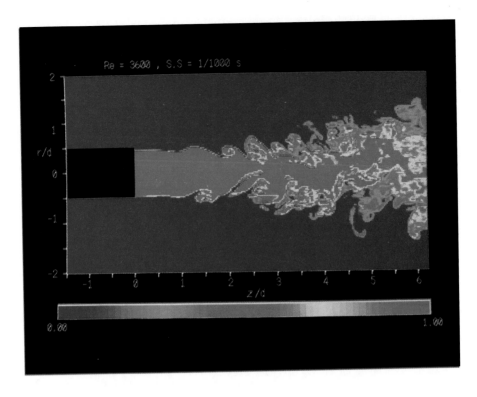

INSTANTANEOUS CONCENTRATION FIELD OF A WATER JET INTO A COFLOWING WATER STREAM (*Re* = 3600, *U/u₀* = 1/15)

Figure 1 The planar laser-induced fluorescent technique was applied to visualize the time-dependent behavior of a coaxial jet to understand the transition and mixing processes in the jet. The injection velocity u_0 is 0.6 m/s, the coflowing stream velocity U is 0.04 m/s, and the inner diameter of the injector is 6 mm. The fluorescent dye (disodium fluorescein) in the jet was excited by the irradiation of 488.0 nm emission from a 1 W Ar++ laser. A laser sheet of 0.50 mm thickness was passed through the jet axis, and the induced fluorescence was recorded by a CCD camera. The shutter speed of the camera was 1/1000 s. The color scale represents the relative concentration normalized by jet fluid concentration.

MANNEQUIN UNDER FIRE

Figure 2 An instrumented mannequin is used for testing the thermal protective qualities of garments when subjected to short duration flash fires. The mannequin is an adult sized male fitted with 110 heat flux sensors distributed approximately uniformly over the body. The sensors absorb energy at rates similar to human skin. The circular tube through the head holds the instrumentation cables. The flash fires are produced with propane diffusion flames. Choked flow orifices are used to control the gas flow through the 12 burner heads which surround the mannequin. The flame motion and structure are a result of gas mixing with the surrounding air and buoyancy induced by the combustion. The average heat flux to the surface of the mannequin is typically 80 kW/m2 for durations up to 5 seconds. The photograph was taken with a single lens reflex 35 mm camera under automatic exposure control. ASA 100 film was used.

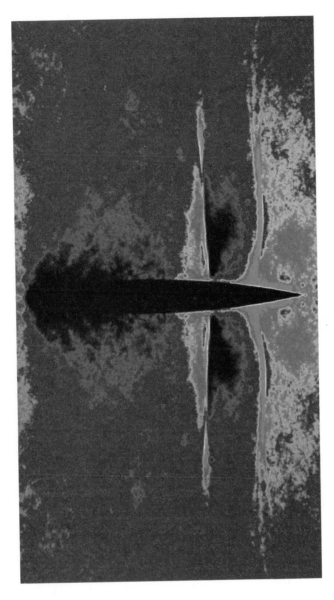

NONEQUILIBRIUM PHASE-TRANSITION IN TRANSONIC FLOW OVER AIRFOIL WING SECTIONS

Figure 3 Homogeneously condensing transonic flow over symmetric airfoil with supercritical heat addition and steady double shock system. The Mach number in the oncoming flow from the left is 0.784. The working fluid is humid air with reservoir conditions, temperature $T_{01} = 295, 6K$, relative humidity $\emptyset_0 = 64, 1\%$ and absolute humidity $x = 11.1\ g_{vapor}/kg_{carrier\,gas}$ (mixture pressure $p_{01} = 1.0\ bar$). In adiabatic flow there is only one shock near the trailing edge. Here an additional normal shock is formed by supercritical heat addition. The flow accelerates from the tip to supersonic ahead of the first shock and it decelerates to subsonic through the first shock. Then, between the two shocks, it accelerates again to supersonic and decelerates to subsonic through the second shock and in the wake flow. Circular arc profile: 10% thickness, chord length 80 mm, Reynolds number $Re = 10^6$. Visualization is by schlieren optical system, knife edge normal to main flow direction, sparc light source with exposure time of 10^{-6} s (reflected picture of half profile experiment). Colors are produced after screening into the computer. Dark blue and violet indicate density decrease; yellow, green, and light blue appear in the diabatic compression region in front of the first shock and behind the second shock.

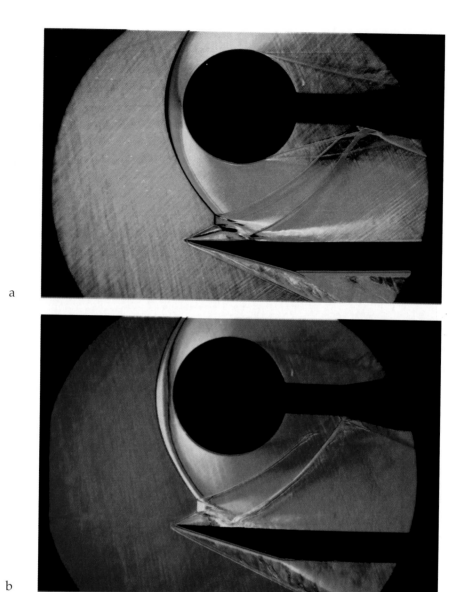

a

b

Color Schlieren Methods in Hypersonic Flow Research

Figure 4 The flow pictures were obtained using the reflected shock tunnel of the Takasago R & D Center. Mitsubishi Heavy Industries Ltd. (a) Shows the shock and boundary layer interaction around a cylinder of 76 mm diameter and flat plate in hypersonic flow ($M_\infty = 4.1$ in air). The direction-indicating-method (a) records all density gradients present in the test section so that the bow shock, the expansion zones, as well as shock interaction with the boundary layer, are represented equally well. The color of the bow shock changes slightly because the direction of the density gradient induced by this shock varies. (b) Shows the magnitude-indicating method. The colors indicate clearly how the density gradient gradually changes from strong compression (light violet) to weak compression (red) to neutral (green) and to weak expansion (blue) to strong expansion (bluish white).

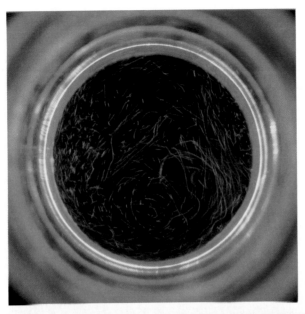

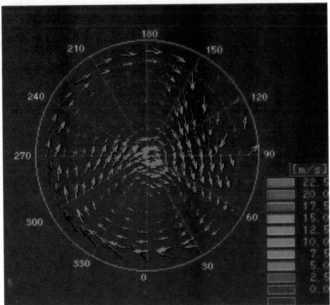

VISUALIZATION OF STEADY-STATE SWIRLING MOTION IN CYLINDER BY MEANS OF TRACER LASER SHEET TECHNIQUE AND 3-DIMENSIONAL LASER DOPPLER VELOCIMETRY FOR DI DIESEL ENGINE

Figure 5 The reduction of exhaust gas emissions is an urgent and an important task for diesel engines. An optimization of in-cylinder air motion represents one of the technological advances towards this objective. These figures show the formation of in-cylinder air swirling motion by using steady state air flow apparatus. Experimental methods used for the visualization and the velocity measurements were a tracer laser sheet technique (TLS) and laser doppler velocimetry (LDV), respectively. The close agreement between the flow pattern of the TLS result and the arrows diagram of the LDV measurement was obtained, especially in the location of separation and vortex centers and vortex numbers. The results from this study show that the air motion in the cylinder mainly generates the swirl motion by a strong vortex from the helical port. It was also found that the swirling flow expands outward from the cylinder center to the cylinder periphery as it goes down through the cylinder.

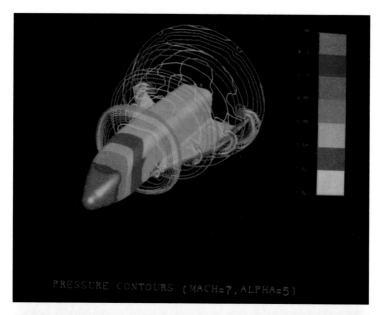

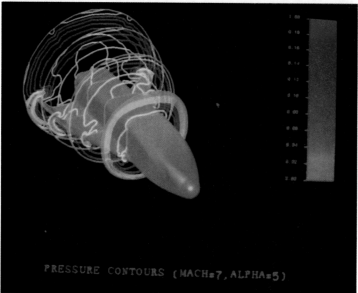

PRESSURE CONTOURS ON THE SPACE SHUTTLE ORBITER

Figure 6 This picture shows the pressure contours of the hypersonic inviscid flow around the HERMES-like space shuttle orbiter. The flowfield has been simulated by the time-dependent algorithm and implicit spatial-marching iteration algorithm. NND scheme (Nonoscillatory, containing no free parameters and dissipative), developed by Prof. Zhang Hanxin in CARDC, is used in the simulation. The spatial grids are generated by parabolic equations mixed with algebraic equations, and the grids of body surface on circumferential direction are distributed. According to the flowfield structure in detail, the freestream condition is M = 7 and the angle of attack is a = 5°.

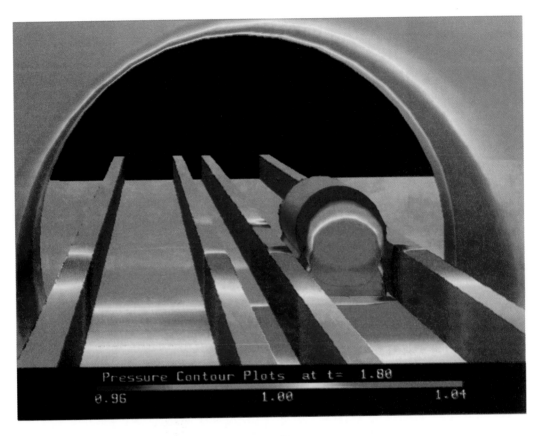

Pressure Contour Plots at t= 1.80
0.96   1.00   1.04

NUMERICAL SIMULATION OF A HIGH-SPEED MAGLER TRAIN MOVING INTO A TUNNEL

Figure 7 Aerodynamics of high-speed trains moving into a tunnel are analyzed with numerical simulation using 3-D Navier-Stokes equations. This picture shows the pressure color maps on a train rushing into a tunnel at 500 km/h. As the color bar says, red corresponds to high pressure and blue corresponds to low pressure. This picture shows the pressure field just after the nose of the train moves into the tunnel, and thus, a rapid pressure increase is observed on the tunnel wall. Since the train is running on the left side, the tunnel wall has higher pressure on the left side and lower pressure on the right side. A so-called guideway, which is the magnetic levitation system, is included in the simulation. Although the local flow field is strongly influenced by the guideway, the global flow field is not influenced as much.

Turbulent Flow Over a Riblet Surface (a)

Figure 8 Three-dimensional particle tracking velocimetry was applied to the measurement in a turbulent water flow in the channel with a riblet surface.* The rib spacing and the height were 3.5 mm and 2.2 mm, respectively. The Reynolds number based on the channel width and the maximum velocity was 6050. The resultant rib spacing, nondimensionalized by the wall variables, is about 15, which would achieve the maximum drag reduction. Neutrally buoyant spherical particles having a diameter of 0.25 mm were used as flow tracers. The measurement volume is about 40 x 40 x 40 mm³. Figure (a) shows a typical spatial distribution of instantaneous velocity vectors measured. About 270 velocity vectors were obtained at this instant. An end view (in the y-z plane) of the conditionally averaged velocity field associated with the ejection (Q2) event is shown in Figure (b). The detection point was located by y at ~ 14, while the threshold factor was given as $H = 1.0$. Note that the reference vector at the bottom of the figure corresponds to 0.9 times the friction velocity. The dimensionless spacing of the grid is 25 in both y and z directions. A quasi-streamwise vortex having a diameter of 60 is clearly seen on the left hand side in the figure, being accompanied by a smaller secondary vortex. The comparison with the result on a smooth wall indicates that the spanwise velocity near the crests of riblets is considerably damped by the riblets.

*Y. Suzuki and N. Kasagi, 1994, Turbulent drag reduction mechanism above a riblet surface, AIAA J., Vol. 32, No. 9, pp. 1781–1790.

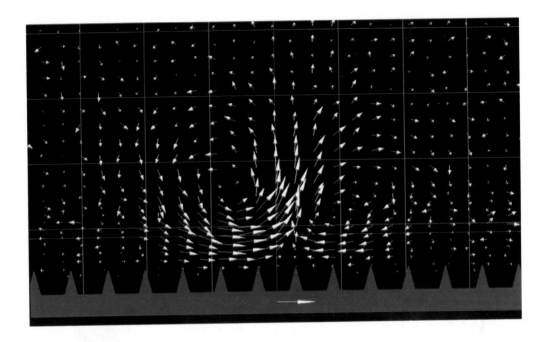

TURBULENT FLOW OVER A RIBLET SURFACE (b)

chapter one

New Visualization and Imaging Techniques for Engine Combustion Research

Tsuyoshi Asanuma

University of Tokyo

Abstract — During the last decade, there has been a drastic change in the nature of engine combustion research. This has been brought about by the close cooperation between the novel laser diagnostics and the large-scale computational facilities, and thus it has led to new visualization and imaging techniques for engine combustion research.

This chapter will review the new techniques in both laser light sheet visualization and computer aided visualization. The former is an optical visualization technique, which makes it possible to visualize the transient phenomena in an engine cylinder by applying the laser-induced diagnostics (e.g., Mie scattering, Rayleigh scattering, fluorescence, and incandescence). In addition, a brief survey of the research engines for optical visualization will also be made. The latter is an innovative image displaying technique which allowed one to plot 2- and 3-dimensional images on the basis of the visualized, measured, or calculated results of in-cylinder events by using fast and rich computer systems.

Introduction

Combustion process in an internal combustion engine is particularly characterized by the following features: (1) engine combustion is usually rapidly changing cycle to cycle during very short periods (e.g., 10 ms at 6000 rpm), (2) it is composed of various transient phenomena such as air motion, fuel mixing, ignition, and flame propagation, etc., (3) it always occurs in a small space enclosed completely by the metal walls of the cylinder head, barrel, and piston crown. Previous investigations, therefore, have mainly focused upon the single-point measurements using intrusive or optical diagnostics together with the conventional visualization techniques, which left a great part of the combustion phenomena in the cycliner to remain unknown, as in a black box.

During the last decade, however, there has been a dramatic change in the nature of engine combustion research. This has been caused by the close collaboration between the new diagnostic approaches based on the nonintrusive laser light sheet (LLS) technique and the more widespread access to large scale computational facilities. The splendid cooperation between both these novel and powerful techniques has led to new visualization and imaging techniques for engine combustion research. One is a high speed and quantitative visualization technique, which makes it possible to visualize more vividly the transient events in the cylinder by applying the pulsed LLS to the research engines for optical visualization, moreover, it yields a more detailed image display with the aid of a computer system. Another technique is an innovative image displaying technique, which allows one to plot 2- or 3-dimensional quantitative images on the basis of the visualized, measured, or calculated results of the in-cylinder events by employing fast and rich computer resources, and thus allows the development of very complex modeling approaches offering the promise of prediction capabilities.

A large majority of the high speed visualization and imaging techniques have always used laser diagnostics in close combination with computer systems. In the following, however, a review will be made of the new visualization and imaging techniques from both aspects of the application of laser light sheet (Laser Sheet Visualization) and the utilization of computational facilities (Computer Aided Visualization), respectively.

Laser Sheet Visualization (Part I, LSV)

Optical Research Engine

To achieve observation of engine combustion in a small chamber closed completely by fixed or moving walls, several research engines designed for optical visualization have been used. At first, a square piston and cylinder assembly with two parallel quartz walls was built to yield full optical access throughout the piston stroke and to observe the side view of the fluid motion and flame propagation by high-speed or schlieren photography (Namazian et al., 1980). The engine was termed "square engine for side viewing." Although side gaskets and piston ring assemblies have been the critical parts for gas sealing, the square engine was used widely to observe the side view of in-cylinder phenomena (Fujikawa et al., 1988). It was, however, clearly confirmed that the behavior of the fluid or flame in the square cylinder was extremely different from that in the circular one particularly for the swirl flow. Hence, it seems to be inappropriate to use the square engine for more realistic studies.

Considering the utilization of laser diagnostics for engine combustion research, Sandia National Laboratories have developed a special optical research engine, as shown in Figure 1 (Smith, 1980). In the engine, the intake, exhaust valves, and spark plugs were located around the side wall of the disc- (Direct Injection Stratified Charge) type combustion chamber which was composed of a flat quartz collection window in the cylinder head and a flat piston crown. Thereafter, the optical engine was named "disc engine for top viewing" and has been frequently used to measure the gas temperature, nitrogen density, turbulent flame structure, and hydro-carbon concentration. Further, a few disc engines with a transparent cylinder head and transparent piston crown termed "disc engine for top and bottom viewings" have been developed for the real time viewing and recording of combustion and related processed in the cylinder by employing a closed-circuit television technique (Steinberger et al., 1979). However, these disc engines also differ remarkably in the configuration of the combustion chamber on comparison with the actual production engines. Therefore, the swirl and turbulence in the intake flow, which strongly affect the combustion performance, are extremely different in both the disc and actual engines.

In order to maintain the actual configurations of intake port and combustion chamber, the most useful engine modification for visualization was an elongated piston with a quartz window in the crown giving a bottom view of in-cylinder events. The engine was called an "actual engine for bottom viewing" (Kozuka et al., 1981). The engine recently combined a pair of quartz windows on the cylinder barrel to allow irradiation of LLS from an outside cylinder, as shown in Figure 2, which was used for the measurements of air-fuel

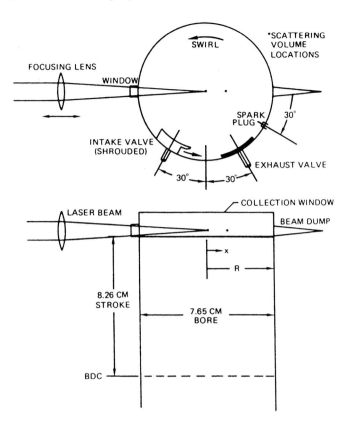

Figure 1 Disc engine for top viewing.

mixture distribution by laser-induced exciplex fluorescence technique (Shimizu et al., 1992). Sometimes the engine included a transparent cylinder to give a side view. Then the "actual engine for side and bottom viewings" was motored with a glass liner and then fired with a sapphire liner for combustion observation by the oil-fog tracer method or the laser-induced phosphorescence technique (Bates, 1988).

To avoid the optical window in these actual engines from being covered with lubricating oil and soot, and furthermore, to continue the engine operation for at least over one hour, the actual engine should be built taking care not to use oil-lubrication and not to

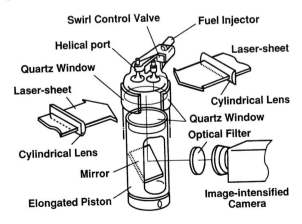

Figure 2 Actual engine for bottom viewing with laser sheet.

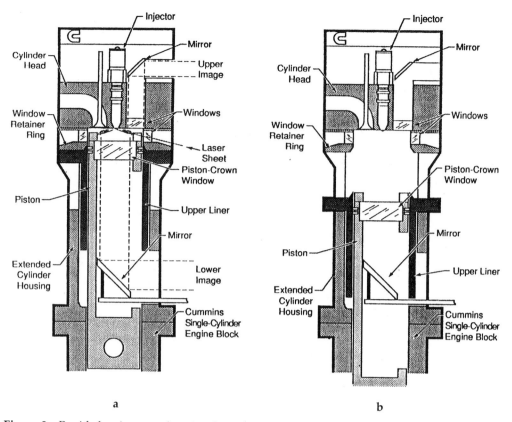

a   b

Figure 3 Rapid cleaning actual engine for top and bottom viewings: (a) operating; (b) cleaning.

overheat the piston crown. Since frequent window cleaning is required for the application of most optical diagnostics to diesel combustion, recently a new optical-access actual engine for bottom viewing has been developed, whose cylinder barrel was composed of extended cylinder housing and upper liner to allow rapid cleaning of all optical windows. When the engine is operating, the upper liner is attached to the window retainer ring as shown in Figure 3a, while the upper liner slides down by removing only two bolts, allowing the optical surfaces to be accessed through two large openings as indicated in Figure 3b. Hence, the entire window cleaning operation can be accomplished in a few minutes, and thus the engine was named "rapid cleaning actual engine for top and bottom viewings" (Epsey and Dec, 1993). These actual engines for bottom viewing will be used more widely for engine combustion research in the future.

Laser Light Sheet (LLS)

Earlier studies on engine combustion have generally used laser beams only as a light source in the conventional optical visualization methods, e.g., shadowgraphy, schlieren photography, and holography, etc., A major limitation of these methods is that they yield information integrated along a line of sight through the in-cylinder field. An alternative approach, which is capable of giving spatially resolved information, is to use LLS illumination to perform 2-D imaging of plane in-cylinder events, such as flow, mixture, spray, flame, and soot, etc. Therefore, the majority of optical visualization with a laser beam may be the "laser sheet visualization," using its scattering from particles or radicals.

Scattering of laser light is usually classified into both elastic and nonelastic groups as listed in Table 1. Among them, the principal scatterings applicable to the laser sheet visualization are counted as follows: Mie scattering, fluorescence, phosphorescence, and

Table 1   Scattering of Laser Light (488 nm)

Kind of scattering		Scattering object	Scattering area (cm/mole/steradian $\times 10^{30}$)
Elastic scattering	Mie	Particle (0.1 ~ 10 μm)	~10^{20}
	Rayleigh	N_2	~10^3
Non-elastic scattering	Raman	N_2	~10^0
	CARS	N_2	~10^7
	Fluorescence	Radical (t < 10 ns)	~10^{15}
	Phosphorescence	Radical (t > 10 ns)	~10^{10}

Rayleigh scattering, according to the order of its scattering area or intensity; whereas CARS and Raman scattering have received more general applications to single-point measurements due to their weak scattering. Although laser-induced incandescence is a kind of thermal radiation and not scattering, it should be described here because it has been noteworthy for soot research in diesel combustion.

LLS can be easily produced with a cylindrical lens or a circular glass rod. The thickness of LLS corresponding to the spatial resolution is usually ranged among several 10 microns and a few mm according to the laser beam diameter. High-power laser sources, Nd-YAG, excimer, and, recently, copper-vapor (CVL), have been employed. Further, image intensified CCD, diode array, or drum camera is generally used to visualize the in-cylinder phenomena illuminated by LLS. For better understanding, this chapter will consider each laser sheet induced scattering technique.

LLS-Induced Mie Scattering Technique (LIM)

The intensity of Mie scattering from seeding particles is the strongest compared to the other scatterings, as shown in Table 1. Hence, Mie scattering has mainly applied to the laser sheet visualization for engine combustion research, whose typical examples will be described.

Air Flow: In a disc engine motored for side and bottom viewings, jet flows during intake process and around valves were observed by using cigarette smoke tracer, a pulsed copper vapor laser sheet, and a streak camera. As shown in Figure 4, the high-speed films provided the insight into the detailed structure of in-cylinder flow field not available from single-point velocity measurements. In Figure 4a, the oscillatory nature of the entering jet flow through the intake valve is visible. The horizontal view (Figure 4b) of in-cyclinder flow on the expansion stroke is applied to estimate the swirl ratio quantitatively (Eaton and Reynolds, 1987). By using copper-vapor pulsed LLS and seeding phenolic microballoons (Phe.MB), high-speed flow visualization was able to image the air flow in a four-valve motored engine with a quartz cylinder sleeve for side viewing. The results indicated that the intake port configuration played a significant role in the generation of initial tumble motion in the early stage of the intake stroke (Lee et al., 1993). In-cylinder flow field in an actual engine for bottom viewing was visualized by seeding expancel particles, irradiating an LLS, and using a CCD camera with an image-intensified gate. Original path-lines of tracer particles were image-processed to make the distinction between particle images and background, and then to obtain both ends of each path-line. Finally it was possible to achieve velocity vector maps of in-cylinder flow. This method, therefore, was able to analyze quantitatively the in-cylinder flow in real time (Fukano et al., 1993).

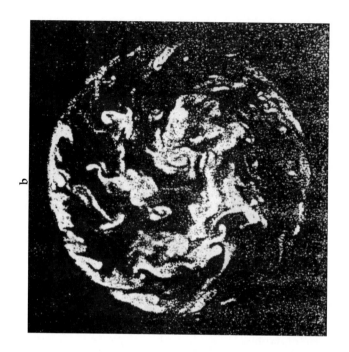

Figure 4 Images of in-cylinder flow at 500 rpm (Mie scattering): (a) vertical view of entering jet flow at 50° ATDC without swirl; (b) horizontal view of in-cylinder flow at 140° ATDC with swirl.

Fuel Spray: To analyze the atomization mechanism of fuel spray injected into a rapid compression engine, the distribution of fuel vapor concentration in an evaporating spray was measured quantitatively by the Mie scattering imaging technique. When fuel containing silicon oil was injected into a high-temperature gaseous nitrogen, the volatile base fuel in the spray vaporized rapidly leaving small droplets of silicon oil. These droplets were illuminated by a thin LLS, and the scattered light was imaged by a still camera to estimate the distribution of vapor concentration in an evaporating spray at various injection pressures and surrounding gas densities (Kosaka et al., 1992). Transient hydrogen free jet ignited near the nozzle exit was observed by the Mie scattering technique. The results revealed that no ambient air is entertained from downstream near the jet tip under the combustion condition, and the air entrainment from the side of the jet in burned or burning gas was smaller than that in an unburned one (Tomita et al., 1993).

The pilot fuel sprays in an optically accessible diesel engine for top viewing were observed by an illuminating pulsed YAG LLS and imaging on a diode array camera to demonstrate repeatability of fuel spray images and evolution of pilot and main spray. As shown in Figure 5, the liquid pilot spray still attaches to the injection tip at 14° BTDC, but detached from the tip at 8° BTDC. The main spray is visible at 4° BTDC, and after TDC the spray is highly affected by the swirl flow (Pushka et al., 1994). The dense core region of diesel spray injected into a test bomb was also investigated by the Mie scattering technique with LLS. Enlarged images of transient spray showed complete atomization and no evidence of an intact liquid core beyond 70 nozzle diameter downstream from the nozzle exit. The image-processed results showed that the illumination emanated from randomly spaced point sources, which were interpreted as droplets (Gülder et al., 1994).

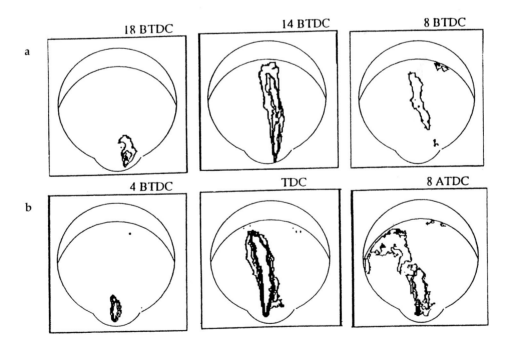

Figure 5   Pilot and main spray images taken at different crank angles and 800 rpm (Mie scattering): (a) pilot spray; (b) main spray.

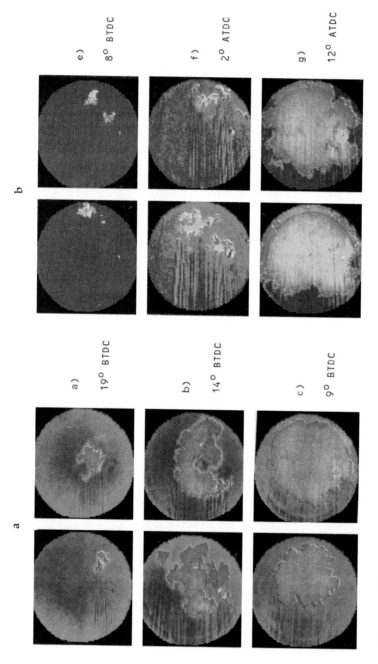

Figure 6   Flame front images taken at 300 and 1200 rpm and indicated crank angles ($\varnothing = 0.6$, Mie scattering): (a) 300 rpm; (b) 1200 rpm.

Flame in SI Engine: Until recently, a square engine for side viewing was employed to perform 2-D flame LLS visualization by Mie scattering from smoke particles. The flame contours were analyzed to determine radii of curvature, wrinkling factors, and turbulent burning velocities at various engine speeds and equivalence ratios (Ziegler et al., 1988). Mie scattering of a pulsed LLS from seeding particles was also used to visualize the details of wrinkled flame fronts in a premixed charge-ported disc engine for top viewing. The leaner flame fronts were repeatable, mostly in size and wrinkling on comparison to the time-sequence of front images taken under the same operating conditions, as can be seen in Figure 6. Furthermore, lean flame fronts for 1200 rpm are distorted more effectively than lower speed ones due to both the intensities of turbulence and the bulk flow in proportion to the engine speed (zur Loye et al., 1987).

In succession to these studies, the first 3-D visualization of flame propagation in the same disc engine was performed with four parallel sheets of pulsed laser lights. The laser sheets (λ = 355, 436, 532, and 683 nm) were arranged with vertical separations between successive sheets ranging from 0.9 to 1.5 mm, and passed through a quartz ring in the cylinder head parallel to the piston top. The lights from the four LLS were scattered by TiO_2 particles and collected by an intensified diode array camera. Some fragmental flames were found in the sets of flame images at lean ($\varnothing$ = 0.59) or higher speed conditions, but were also supposed to be continuous in a vertical direction. The simultaneous flame images on four different planes, as shown in Figure 7, confirmed that the flame surface was composed of contiguous flame sheets and no island of products or reactants was observed in any flame regimes (Mantzaras et al., 1988). On the other hand, the structure of premixed turbulent flames in a constant-volume bomb was visualized using LLS-induced Mie scattering from fine particles to TiO_2. According to the 2-D flame images, islands of reactants as well as of products were found to exist when the turbulence intensity was above 0.4 m/s. In contrast to the previous results, this may be presumed due to the extreme difference in the combustion chamber shape and the operating conditions (Kido, H. et al., 1993).

Soot in Diesel Combustion: Mie scattering LLS images of a soot cloud formed during combustion were obtained at different cross-sections in the deep bowl of a disc engine for top viewing. Comparing these images with direct flame photographs taken simultaneously, it was revealed the soot cloud was mainly formed either in the periphery of the flame or in the vicinity of the chamber wall, where hot over-rich gases may be stagnant and quenched (Shioji et al., 1992). Similarly, two soot particle images in an unsteady combusting spray injected into a quiescent atmosphere were taken by using double-pulsed LLS and two intensified gated cameras. The Mie scattering images of soot in combusting spray for various injection pressures clarified that soot was mainly formed in the periphery of the flame tip, as can be seen in Figure 8, where the air entertainment was less and flame temperature favored the soot formation (Won et al., 1992). To visualize the flame and the soot clouds formed in and around the piston cavity, LLS Mie scattering was used. The soot clouds could be discriminated by the CCD images near the cavity wall and at the exit of the cavity during the later stage of combustion (Ota et al., 1993).

LLS-Induced Rayleigh Scattering Technique (LIR)

Hitherto, laser Rayleigh scattering has applied entirely to the single-point measurement of fuel vapor concentration near the spark plug (Zhao et al., 1991). This method was recently improved to obtain the 2-D distribution of the mixture or gas jet. The absence of droplets

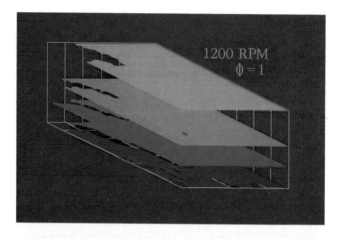

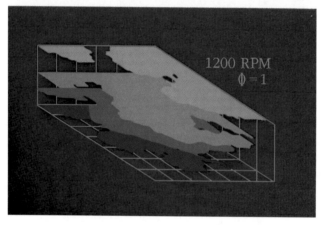

Figure 7   3-D representations of stoichiometric flames visualized with 4-LLS simultaneously (Mie-scattering).

is, however, fundamental for the application of Rayleigh scattering to the fuel vapor, because Mie scattering from fuel droplets and dirty particles causes a strong signal interference.

A propane-fueled SI actual engine motored for bottom viewing was employed to acquire the 2-D images of mixture density distribution using LLS-induced Rayleigh scattering (LIR) technique. As a result, the effects of fuel injection timing, injection direction, and in-cylinder air flow on the mixture distribution have been clarified (Zhao et al., 1993). The same technique was used to visualize the concentration field of gas jets discharged into a dust-free air stream in the atmosphere, and then the temporally frozen images were quantified with the image analysis system. The results showed that the large eddy structure occurred at the tip of the unsteady jet and the concentration profile resembled a mushroom (Tsue et al., 1993). The LIR technique was also applied to the rapid cleaning actual engine for top and bottom viewings (see Figure 3) to perform quantitative imaging of the fuel vapor concentration before ignition. The spatial sequences of the equivalence ratio (Ø) in the vapor fuel region near the leading edge of the diesel spray jet at 4.5° after the start of injection are shown in Figure 9. The results in this vertical sequence supported that the fuel and air mixed relatively well throughout the leading portion of the jet, and the mixing proceeds so rapidly that the equivalence ratio in the leading portion decreases in the range of Ø = 1 ~ 2 during only half degrees CA (Espey et al., 1994).

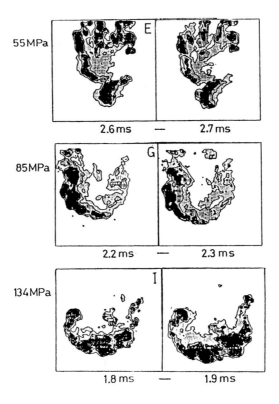

Figure 8 2-D soot images of combusting fuel spray for various injection pressures and time (ms) lapses from injection start (Mie scattering).

LLS-Induced Fluorescence Technique (LIF)

A matter which has absorbed an amount of energy is, generally, apt to release a part of its energy as a visible ray or a magnetic wave with a similar wavelength. Such released rays

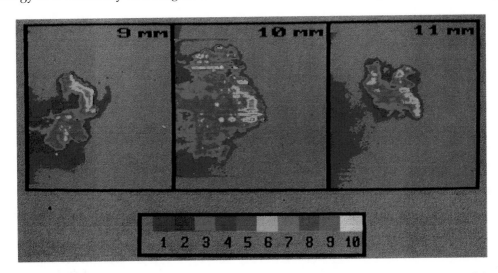

Figure 9 Spatial sequence of equivalence ratio in vapor-fuel region near the leading edge of diesel spray jet at 7° BTDC and 1200 rpm. Distance from cylinder head to image plane (LLS) is given in upper right of each image (Rayleigh scattering).

are termed "luminescence." Among them, ones with illumination life shorter than 10 ns are named "fluorescence," while those with a longer life are called "phosphorescence," as listed previously in Table 1. Since the former was found to give an excellent scattering intensity, it is widely used as an LIF technique, including the following two ways: one is a method capable of observing the flow field by radiating the seeded fluorescent particles (Iodine, I_2) with a laser light of a fixed wave-length, and accordingly, it is used mainly in the field of optical flow visualization; another method is to visualize the in-cylinder events by the fluorescent ray from the radicals (CH, OH, O, and NO) of the working fluids or seeding particles (biacetyl), when a continuous or pulsed laser with variable wave-length is in reasonance with them. Therefore, the latter is principally applied to engine combustion research.

Gas Flow: In-cylinder flow seeded with 2 ~ 3 μm phosphorescence particles was visualized by a single-pulsed UV laser and digitized for quantitative analysis (Bates, 1988). The mixing process of intermittent gas jets discharged into a test bomb also was visualized using iodine in the ambient gas excited by LLS. The LIF images showed that a considerable amount of air was entrained just under the umbrella profile at the top of the unsteady jet. In addition, the mean concentration of the gas jet was found to relate to only one parameter, that is a ratio of the discharge Reynolds number (Re) and the nondimensional length (= jet nozzle diameter/jet length) (Kido, A. et al., 1993). Similarly, the LIF from iodine seeded into a transient gas jet was used to get the instantaneous density images of the jet flow. On the basis of the 2-D images, the effects of several factors, such as gas concentration, temperature, and kind of ambient gas on the fluorescence intensity were examined to establish a quantitative LIF technique for density distribution in a gas jet (Ando and Iida, 1993).

Mixture in Engine: The LIF technique was used to obtain 2-D images of molecular density distribution in a modified disc engine for bottom viewing. The fuel, iso-octane, was visualized in the intake of mixture, in the ignition, and in the movement of flame front, as shown in Figure 10. The figure is at 340° CA for the initial fuel-air intake. Later, in the induction (370°) and compression (650°) phases, the mixture distribution becomes more homogeneous before combustion. The spark plug fires at −6°, and these figures at 10 and 30° show that there is still some fluorescence behind the flame front, which indicates some unburned fuel (Andresen et al., 1990). Ethylmethyl-ketone (EMK), similar to gasoline in boiling temperature and vapor pressure, was injected into the intake manifold in a transparent SI square engine for

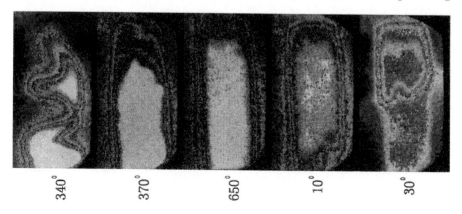

340°   370°   650°   10°   30°

Figure 10   2-D images of iso-octane mixture concentration distribution before and after combustion spark-ignited at 714° (or −6°) (LIF).

displaying LIF images of fuel distributions in a combustion chamber during the intake and compression strokes at 1000 rpm. From the fluorescence signals, 2-D images of air/fuel ratio were able to be estimated using calibration data from bomb experiments (Lawrenz et al., 1992). A dopant, biacetyl, with vaporizing properties equivalent to those of the gasoline was used as a fluorescent tracer to acquire the LIF quantitative images of the mixture in an optical access engine for bottom viewing at 1200 rpm. The LIF technique was able to quantify the mixture homogeneity. The 2-D LIF images showed that the mixing was far from being perfect at BDC, but tended to homogenize during compression and no droplets were observed at TDC (Baritaud et al., 1992). Similarly the LIF technique used fluorescence from hydrocarbon molecules to image the fuel vapor distribution in a disc SI engine for side and bottom viewings. Here it is noteworthy to recognize the inhomogeneous distribution of fuel throughout the long periods of intake stroke, and particularly, the high fuel concentration found below the inlet valve (Winklhofer et al., 1993). The cyclic variation of the mixture concentration was also measured in an SI engine for bottom viewing with the application of 2-D LIF technique. As a result, the random distributions of the heterogeneous mixture with different size lumps were presumed to play an important role in determining the cyclic variation. In addition, the mixture heterogeneity becomes smaller and its cyclic variation decreases with the lapse of the engine crank angle and the increase in engine speed (Zhao et al., 1994). Further, 2-D mixture images in an actual SI engine for bottom viewing have been obtained by using LIF technique with acetone as the tracer. For different engine parameters, i.e., injection timing, equivalence ratio, EGR, etc., both snap shots (single cycle shot images, Figure 11a, and averaged images, Figure 11b) of fuel distributions were detected at various crank angles during the intake and compression strokes. Both images of the first row (524°) in Figure 11 demonstrate that the big fuel droplets inside the cylinder are originating from fuel films on the wall of intake valve and port. The images of the third row (4° CA) in the figure show nonuniformity of the fuel distribution after ignition (Wolff et al., 1994).

Flame in Engine: Recently, the planar imaging of LIF from OH radicals has been employed to visualize 2-D flame contours and to identify the location and shape of the propagating flame front. The OH LIF results were roughly similar to the corresponding Mie scattering results which rely on doping the combustion gases with particles as indirect markers of the flame, while LIF provided a direct marker of a species (OH) naturally associated with the combustion process. The OH fluorescence images from a flame in a disc engine for top viewing were easily acquired, because the fluorescence rose sharply through the flame front to a peak value and then decreased gradually in the post-flame gases (Felton et al., 1988). Similarly the LIF technique was applied to visualization of OH radical distribution in an actual engine for bottom viewing under various operating conditions, e.g., mass burned fraction, swirl ratio, and air-fuel ratio. As shown in Figure 12, the flames in the direct images spread with a smooth flame front, while the flames in the fluorescence images propagate with complex peninsular shape. When comparing both images, the OH radical was found to be mainly generated at the higher temperature reaction zone in the post-flame gases (Tanaka and Tabata, 1994). To clarify the effect of the spark gap width, the spark ignition process of a methanol-air mixture in a combustion chamber has been studied 2-D LIF imaging of the OH radical. The OH images revealed that just after the onset of spark, the flame kernel is compressed around the electrode axis, and if the spark energy is large enough, a self-sustained flame will propagate from the surface of the toroidal ignition kernel into the surrounding combustible mixture (Xu et al., 1994).

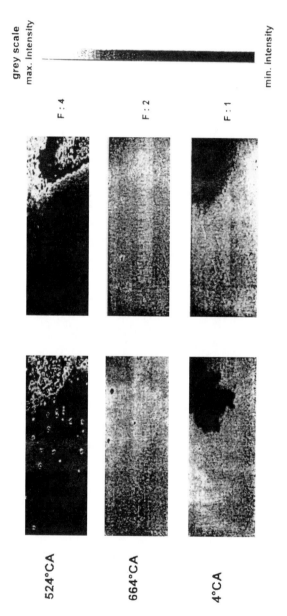

Figure 11 In-cylinder images of acetone mixture concentration after port-fuel injection during intake (524) and compression (664) strokes at 1400 rpm and ∅ = 1.0 (LIF): (a) single cycle distribution; (b) averaged distribution.

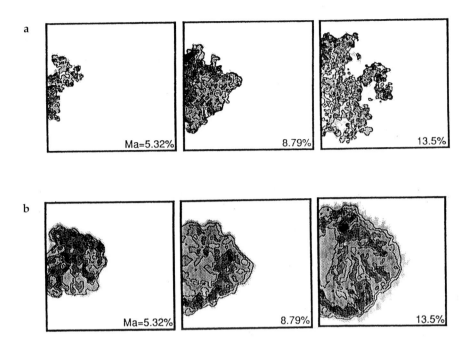

Figure 12   Simultaneous time history images for A/F = 14.7 (Ma = Mass burned fraction): (a) LIF images of OH radicals; (b) direct images of flame.

On the other hand, LIF imaging was the most suitable approach for characterizing the NO evolution process in the square combustion chamber of a disc diesel engine for top viewing. The greatest obstacle to acquiring NO images was the attenuation of the laser beam by particulate emissions. Hence, oxygen was added to the intake air charge to reduce the emissions in the combustion medium. Then the NO images were of sufficient clarity to resolve the structure of the NO formation process, which was found to start almost immediately after ignition and to cease by 30 ~ 40° ATDC (Alatas et al., 1993). It is generally accepted that autoignition in the SI engine starts in distinct areas of the end gas, which is caused by inhomogeneities in temperature and/or in charge distribution. Planar LIF technique was also applied to a disc engine for top viewing at 1000 rpm. The LIF images of formaldehyde were taken in both knocking and nonknocking cycles, indicating the formation and consumption of formaldehyde in the end gas. It was clarified that autoignition started in clearly bounded regions in the end gas, which were identified by the absence of formaldehyde due to its consumption during the ignition process (Hoffman et al., 1994). In a manner similar to the mixture imaging, 2-D LIF images of a flame in a flat-head engine for bottom viewing were obtained by uniformly seeding a propane-air mixture with biacetyl as a fluorescent tracer. Unlike the Mie scattering technique, it was possible to observe an extremely convoluted flame front due to the very low noise level of the technique. This good definition of the flame front allows one to easily perform quantitative analysis of the flame images on large sampling sets (Baritaud, 1994).

LLS-Induced Exciplex Fluorescence Technique (LIEF)

Recently, it has been shown that spectrally separated fluorescence emissions from the liquid and vapor phases can be obtained by adding exciplex-forming dopants to the engine

fuel and irradiating it with a powerful LLS (Melton, 1983). Consequently, the LIEF technique has been established as a useful and reliable qualitative visualization tool. An
obvious extension of the technique is to calibrate it to make a quantitative measurement of
fuel concentration in both liquid and vapor phases. Accurate quantitative 2-D information
is desirable and important, not only for analyzing in more detail the effect of mixture or
spray on the combustion process, but also for direct comparison with the results of
computational models. Typical examples of LIEF will be given as follows:

Mixture in SI Engine: To apply the LIEF technique to a gasoline engine, the exciplex-
forming dopant must have a boiling point within the distillation range of gasoline
(20 ~ 215° CA). Despite extensive research, the only dopant found to satisfy this
requirement was N, N-dimethylaniline (DMA), whose normal boiling point is
193°C. Figure 13 shows the liquid and vapor phase fluorescence spectra of the 5%
DMA/5% naphthalene/90% gasoline by weight. In the vapor phase, DMA accounts
for about 85% of the fluorescence and is thus the dominant emitter at 335 nm, but
in the liquid phase, this molecule forms an exciplex with naphthalene and the
emission is red-shifted to 410 nm. These dual laser sheets were introduced into an
actual engine for bottom viewing (see Figure 2) to simultaneously record 2-D
images of both the liquid and vapor phases in the mixture during the induction and
compression strokes. The spatial distributions of both phases are quantitatively
compared in Figure 14, demonstrating the effect of swirl flow created by the swirl
control valve on the mixture formation (Shimizu et al., 1992). The same LIEF
technique was also applied to an actual engine for side and bottom viewings to
obtain the density distributions of liquid and vapor phases in a mixed fuel. Both
images of liquid and vapor during the intake and compression strokes were acquired through both band-filters (480 nm for liquid phase and 400 nm for gas
phase), and then used to examine the effect of various swirl conditions on the
mixture formation by comparison with the in-cylinder flow vector maps image-
processed with the particle image velocimetry (see PIV on page 22). Under a low
swirling condition, a rich fuel vapor concentrated mainly at the exhaust valve side.

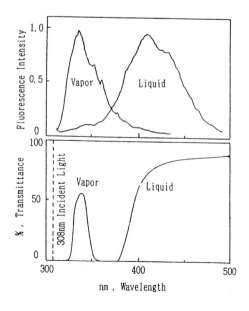

Figure 13   Fluorescence spectra of vapor and liquid phases, and transmissibilities of the filters used.

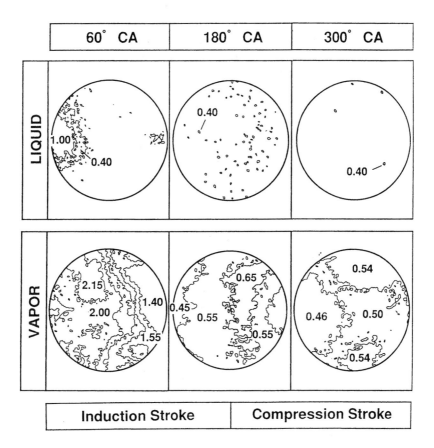

Figure 14 Spatial distributions of liquid and vapor phases at the opening of swirl control valve (LIEF).

With increased swirl ratio, the fuel evaporation rate increased and impinged fuel was reduced, leading to an increase in combustion stability and to a reduction in HC emission (Fujimoto and Tanabe, 1993).

Fuel Spray in Diesel Engine: The LIEF technique was successful for making simultaneous 2-D images of liquid and vapor (gas) density distributions within the fuel spray. The mixed fuel containing exciplex-forming dopants was injected into a disc engine motored for top viewing. These LLS fluorescence signals from liquid and gas phases were separately collected on a diode array camera to acquire both phase images of fuel spray. In Figure 15A, a pair of liquid and gas phase images obtained simultaneously are compared to clarify the evaporation process of the fuel spray. The remains of the core, together with the liquid fuel on and near the wall at the lower right can be seen, whereas gas is observed around the core and spreading mostly to the right with the swirl, but also to the left. In addition, it is necessary to run this in a nitrogen environment to avoid quenching the gas phase fluorescence by oxygen. Hence, some tests were made to determine the quantitative degree of this quenching. Figure 15B shows the effect of oxygen quenching by comparing both images of the gas phase signal in nitrogen and in air. In nitrogen, the contour interval is 2 and the maximum intensity is 9.4; while in air, the interval is 0.05 and the maximum intensity is 0.18. The quenching by oxygen, therefore, reduces the gas fluorescence by a factor of at least 50 at the cylinder pressure near TDC (Bardsley et al., 1988). To acquire more detailed information on the structure of a diesel spray,

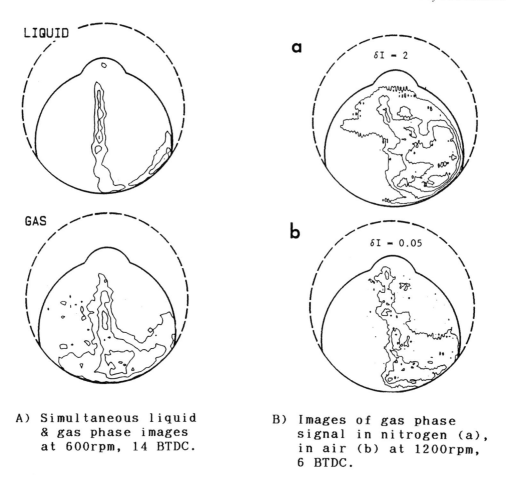

A) Simultaneous liquid
   & gas phase images
   at 600rpm, 14 BTDC.

B) Images of gas phase
   signal in nitrogen (a),
   in air (b) at 1200rpm,
   6 BTDC.

Figure 15   Comparisons of fuel spray images (LIEF).

the LIEF technique was applied to 2-D visualization of a fuel spray discharged into a pressurized, gaseous nitrogen inside an optical bomb. The bomb was previously heated by combustion of a homogeneous mixture of methane and air. Intensity contours of liquid concentration within the spray of the mixed fuel were obtained under both nonevaporating and evaporating conditions (Tsue et al., 1992). By using the LIEF technique, clear 2-D images of vapor and liquid phases in an unsteady single spray impinging on a flat wall were acquired at the same time. It was found that these images corresponded to the predicted results (Kobayashi et al., 1993).

The LIEF technique, moreover, was used to provide information on the behavior of both phases (liquid and gas) of fuel spray injected into the nitrogen gas in a DI diesel engine for bottom viewing. At the beginning of spray penetration, the liquid and vapor advance at the same speed. Soon, the liquid phase stops its progression, while the gas phase continues to penetrate almost the whole chamber (Baritaud et al., 1994). To prove the quantitative applicability of the LIEF to a firing engine, both the thermal decomposition of a TMPD dopant (NNN'N'-tetramethyl-p-phenylene di-amine) and the relationship between fluorescence intensity and vapor concentration of TMPD were examined with a rapid compression machine for top viewing. Consequently, there was no effect of the TMPD thermal decomposition on the vapor fluorescence intensity during a short period (less than 10 ms) under an ambient condition and furthermore, the fluorescence intensity was found to be nearly proportional to the TMPD vapor concentration at a fixed temperature (Yeh et al., 1993).

LLS-Induced Incandescence Technique (LII)

Past attempts at measuring in-cylinder soot distribution had severe limitations for providing an accurate picture of where and when soot occurs in the diesel engine cyclinder, whereas laser-based planar imaging diagnostics have the potential to provide better temporally and spatially resolved measurements of the soot distribution. Among them, LII has developed into a technique for producing planar images of soot particles along in a diesel cylinder. Advantages of LII soot imaging over Mie scattering include no interference signals from droplets, easy rejection of laser light scattered by in-cylinder surfaces, and signal intensity proportional to the soot volume fraction. Conversely, disadvantages of LII imaging are the difficulty of suppressing background luminosity, lower signal strength, and being partially intrusive.

At first, a disc diesel engine for top viewing was operated with low-sooting fuel, i.e., a blend of 80% 2,ethoxyethyl ether and 20% hexadecane, and then an LLS was used to heat soot from the flame temperature (~2200 K) to approximately the soot vaporization temperature (~4500 K) during an 8 ns laser pulse. By collecting the thermal radiation from the heated soot, 2-D images of the soot distribution within a combusting spray plume were obtained through a small diameter window in the cylinder head. The LII images revealed that the soot was distributed throughout the observed cross-section of the combusting spray plume, as shown in Figure 16, rather than just around the periphery as was previously thought from the Mie scattering soot images (Dec et al., 1991). The 2-D LII soot images disclosed that soot distributed not only throughout the combusting region over a wide speed range, but its concentration also decreased as engine speed and injection pressure increased. In addition, both Mie scattering liquid fuel images and LIF vapor fuel images visualized simultaneously were compared with the LII soot images to examine the soot formation process in the combusting spray (Dec and Espy, 1992).

Recently, temporal sequences of flame and soot images were acquired by applying both imaging techniques of natural flame luminosity and LII to a newly designed rapid cleaning engine (see Figure 3). The images showed that the first soot was detectable at the peak heat release rate, while the distribution of soot within the plume did not change greatly until TDC. Once the luminous flame was established, the LII soot images revealed

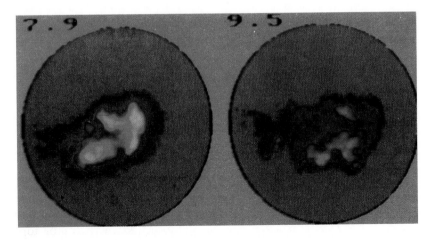

Figure 16   Spatial sequence of soot images taken at 9° ATDC and 600 rpm. Number in each image denotes distance in mm from cylinder head to image plane (LII).

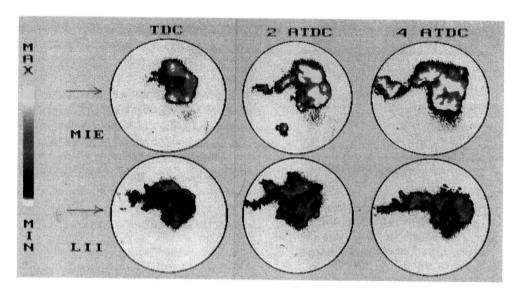

Figure 17   Comparison between Mie scattering (top) and LII (bottom) soot images in combusting spray at low fuel load and 1200 rpm.

that soot was distributed throughout the cross-section of the combusting fuel jet, and the extent of the soot distribution compared well with the extent of the luminous flame region. Figure 17 presents temporal sequences of LII and Mie scatter soot images. They provide different information about the soot distribution, because the Mie scatter signal intensity is proportional to the sixth power of soot particles diameter (d^6), while the LII signal intensity is proportional to the soot concentration (d^3). In the figure, the Mie image at TDC shows a strong signal near the leading edge of the combusting fuel jet and no detectable signal upstream. Conversely, the LII image shows a somewhat weaker, but still easily detectable signal in the upstream region. After injection ended (2 or 4° ATDC), both soot images looked similar. Consequently, soot was found to distribute throughout the cross-section of the combusting fuel jet for all conditions investigated (Espey and Dec, 1993). The above planar imaging technique was also used to clarify the evolution of NO (via LIF) and soot (via Mie and LII) in a diesel engine equipped with a square combustion chamber for top viewing and skip-fired at 900 rpm. As a result, the NO formation starts early during combustion and stops no later than 30 ~ 40° ATDC, while the soot concentration starts increasing at 20° ATDC, and the growth in soot particle diameter can also be observed by comparing the LII and Mie scattering images (Alatas et al., 1993).

Conclusions for Laser Sheet Visualization

By applying the advanced laser diagnostics to the newly developed optical research engines and receiving the aid of computational facilities, the laser sheet visualization techniques are now more available in the various fields of engine combustion research, as described previously. Among them, typical trends will be summarized as follows:

1. To achieve the purpose of laser sheet visualization, "actual engine for bottom viewing" is now employed universally due to its actual configuration of the combustion chamber and its ability to observe in-cylinder events entirely through a piston window. Above all, an innovative rapid cleaning actual engine is also newly developed for soot visualization in diesel combustion, and it may be used widely in the future.

2. Because of high scattering intensity, LLS-induced Mie scattering was mainly used for 2-D imaging of in-cylinder events, e.g., air flow, fuel spray, flame, and soot. Above all, 3-D visualization of premixed charge flame is performed by Mie scattering with 4 parallel LLS. The simultaneous flame images confirmed that no island of products or reactants was observed in any flame regimes. Rayleigh scattering was also useful in yielding the 2-D concentration distribution of gaseous mixture.

3. LLS-induced fluorescence has allowed visualization of the gaseous jet flow seeded with fluorescent particles or the in-cyclinder mixture and/or flame causing resonance with their radicals. Recently, by using spectrally separated fluorescent signals from liquid and vapor phases of fuel added with dopants, LLS-induced exciplex fluorescence technique has succeeded in imaging the concentration distributions of both phases of fuel to acquire quantitative 2-D information about the mixture or spray.

4. Among the past attempts at measuring engine soot distribution, LLS-induced incandescence has proven to be an adequate technique for producing planar images of soot particles alone in a diesel cylinder by collecting the thermal radiation from the heated soot. The images revealed that the soot was distributed throughout the observed cross-section of the combusting spray plume, rather than just around the periphery as was previously thought.

Computer Aided Visualization (Part II, CAV)

Classification of CAV

For engine combustion research, CAV was recently promoted more actively as a new image displaying technique, which allows one to acquire multidimensional quantitative images on the basis of the tremendous visualized, measured, or computed results of the in-cylinder events by employing fast and rich computer facilities. In general, the CAV may be classified into the following three areas:

1. Image Processing of Visualized Photographs: Qualitative photographs visualized by the conventional or LLS technique can be converted to the quantitative 2-D or 3-D images using digital procession and analysis with computational resources.

2. Image Displaying of Measured Data: Vast quantities of point-measurement data, which hitherto would be difficult or impossible to comprehend as it was, interpreted well by image displaying in computer facilities.

3. Image Displaying of Computational Simulation: Extensive results computed numerically with simulation models also require image displaying for comparison with the visualized or measured results for better understanding and validity of the models.

Image Processing of Visualized Photographs

High-energy pulsed laser illumination and digital processing permit the tracer methods in conventional flow visualization to use microscopic seeding particles following the rapid velocity fluctuation in in-cylinder flows. The image processing of visualized photographs is applicable to both fields of velocity vector and temperature distribution. The former further involves the following two groups, particle tracking velocimetry (PTV) and particle image velocimetry (PIV), both of which can provide quantitative velocity vector maps from the path-lines of tracer particles seeded in the flow field. The latter is a photograph processing technique, which can estimate the flame temperature or soot distribution in the combusting flames by image-processing the original photographs of two-wavelengths (hence called a two-color method).

Particle Tracking Velocimetry (PTV)

In the case of PTV, the light source repeatedly pulsed to record particle tracks using various techniques; e.g., continuous laser sheet illumination combined with a multishutter camera, an electro-optic device, or an acousto-optic device. The period of succeeding light pulses (T) must be chosen to permit separation of a single pulse-line of a tracer particle, and the pulse length (t) must be large enough to distinguish succeeding pulse-lines onto the film. After all, a pulse-line with a ratio (T/t) of 5 to 1 has been found suitable for applications of PTV for in-cylinder flows.

A 2-D flow field in a model engine motored for side viewing was visualized with a laser sheet and an acousto-optic modulator driven at 40 MHz. Then the broken path-lines in an original photograph (Figure 18a) were analyzed by using the PTV technique to achieve velocity diagrams, as illustrated in Figure 18b (Hentschel and Stoffregen, 1987). In PTV, however, the particle seeding density must be low enough to obviously identify individual broken path-lines. Accordingly, PTV has recently offered its seat to the next particle image velocimetry (PIV).

Particle Image Velocimetry (PIV)

Auto-Correlation PIV: Only short laser double-pulses were used in PIV to record a double-exposed image for each particle on the film. Hence, the seeding density could increase so sufficiently that PIV was expected to provide a more accurate measure of velocity gradients than PTV. In the past, PIV technique has been employed only to steady water flow fields at relatively low velocities. Recently, the PIV was first applied to a disc motored engine equipped with a pair of small windows for LLS irradiation or LDV measurement. The seeded fine particles (TiO_2) were illuminated by a double-pulsed LLS (20 ~ 40 μs pulse separation) oriented parallel to the piston head. The 2-D velocity vector maps over a 12 mm × 32 mm area were acquired by using the auto-correlation PIV technique, and they have, furthermore, allowed for the evaluation of instantaneous velocity, high-pass filtered velocity, and large scale vorticity with 1 mm spatial resolution, as shown in Figure 19 (Reuss et al., 1989). These data, however, are too limited to offer general conclusions about in-cylinder flow because the interrogated area is much smaller than the cylinder size.

In succession, the PIV was used to make instantaneous velocity measurements over a 24 mm × 32 mm area in a two-stroke disc engine fired for top viewing. Boron nitride particles (0.3 ~ 0.7 μm) were seeded instead of TiO_2 to provide good quality images in the burned gas regions. Nd-YAG laser (532 nm) was double-pulsed with a pulse separation of 35 μs and an individual pulse duration of 10 ns. From the pulsed LLS of 200 μm thick and 50 mm wide, distributions of velocity vector, high-pass filtered velocity, large scale vorticity, and shear-strain-rate in the unburned gas ahead of and near the flame were obtained. The vorticity distribution was superimposed with the high-pass filtered velocity distribution in Figure 20. The figure shows the strong correspondence between the small scale structures in the vorticity and velocity distributions. In particular, the high intensity positive and negative (red and blue) vorticity regions properly identify the counter-clockwise and clockwise vortex structures seen in the filtered velocity distribution. As can be seen in the figure, the attempts to interrogate the burned gas regions (marked by A) have failed because of poor signal-to-noise due to less particle density and because of small and random displacement of seeding particles (Reusse et al., 1990).

Using a movie version of PIV, 2-D time resolved (1000 frames/s) flow field measurements have been made in a square SI engine for side viewing in a plane through the

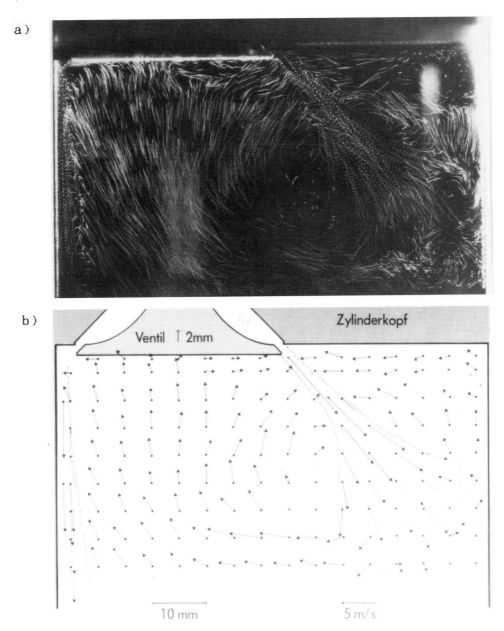

Figure 18 Velocity vector diagram (b) of in-cylinder flow achieved by evaluation of original photograph (a) (PTV).

axis of the inlet valve. This technique was successfully applied to cycle-resolved measurements of the in-cylinder flows using a copper-vapor LLS and a rotating drum camera. Taking all possible improvements into account, it was possible to obtain a spatial resolution of 1 mm in a field of view of 75 mm × 80 mm (Stolz et al., 1992). To investigate the instantaneous 2-D velocity field around the inlet valve by PIV technique, a simulated engine model of an acrylic cylinder with glass windows was used. The double exposed photographic images were taken by using LLS from YAG and seeding TiO_2 particles (1 μm diameter). According to the images, the intake jet thickness became thinner as it moved toward the valve edge. After leaving the valve, the jet interacted and mixed with ambient air and diffused as shown in

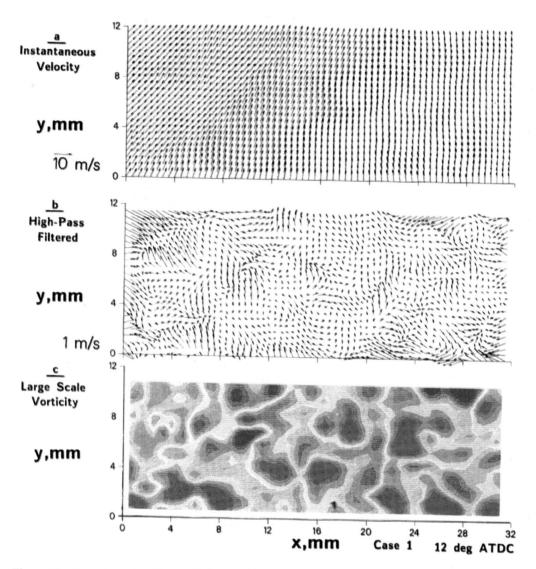

Figure 19   Instantaneous velocity, high-pass filtered velocity, and large scale vorticity distributions of in-cylinder flow at 600 rpm (PIV).

Figure 21 (Lee and Farrell, 1993). PIV has also been used to characterize barrel swirl (tumble) throughout the compression stroke in a four valve pentroof optical engine equipped with a glass piston and cylinder. By illuminating with double pulsed LLS and seeding with a fine mist of olive oil, PIV images have been recorded in a plane parallel to the cylinder axis of an IC engine motored at 1000 rpm. The images were interrogated within 1.1 mm square regions with 50% overlap, using a commercial digital auto-correlation system (Reeves et al., 1994). To measure gas velocity and turbulence in the piston bowl of a diesel engine motored at 750 rpm, PIV was employed. The data were analyzed by a fully optical interrogation system with 0.7 mm spot size. Diagrams of PIV velocity vector superimposed with KIVA predicted images have distinctly indicated the organized swirl motion. In addition, both results showed good agreement qualitatively, but the predicted velocities fall considerably short of the measured results (Sweetland and Reitz, 1994).

Color Sheet PIV: To eliminate the directional ambiguity on the position of seeding particles visualized in the flow field, the particles were illuminated with two laser

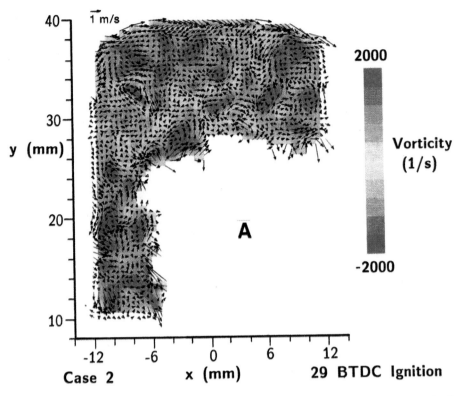

Figure 20 Vorticity (color) and high-pass filtered velocity vector distributions of in-cylinder flow with flame (A) at 600 rpm (PIV).

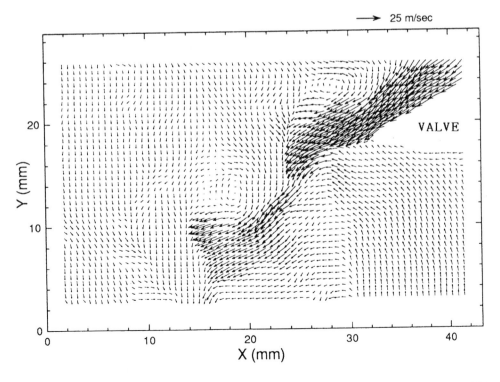

Figure 21 Instantaneous velocity vector of dynamic intake valve flow at 300 rpm cam speed (PIV).

sheets of different wavelengths and were recorded as the temporal images of the path-line with different color ends. This technique is called "color sheet PIV." In-cylinder flow field in a disc motored engine was visualized with two LLS of 532 and 650 nm in wavelength. Since the sequence of the different color images was known, it was possible to remove directional ambiguity. Therefore the color sheet PIV technique was available for studying the complex recirculating flows in an IC engine. In addition, both images of ensemble-averaged velocity and velocity fluctuation determined by PIV were found to compare well with those made by point-measurement of LDV (Nino et al., 1992). The same optical engine with a cylindrical cup-in-head combustion chamber and a spark plug mounted at the center of the piston crown was employed to obtain a 2-D velocity vector map using the color sheet PIV. By separately interrogating the two images of different color, good quality PIV data were obtained. The velocity information in the burned gases, however, was erroneous due to the low seed density and the obstruction of combustion light (Nino et al., 1993). The multi-color PIV was further improved by simultaneously using two- or three-color laser sheets to precisely yield both path-line ends of each particle, and thereby easily distinguish the flow direction. The principle of color sheet PIV is illustrated in Figure 22a. A laser beam is separated into green and blue beams and converted into pulsed laser sheets. The photographs of particles passing through these sheets are composed of green, cyan (green + blue), and blue images. Hence, it was possible to determine the length and direction of the particle path-line. A velocity vector diagram of in-cylinder flow at TDC revealed that the distortion of tumble at 15° BTDC has induced horizontal vortices besides the vertical vortex, as shown in Figure 22b, which resulted in optimization of the combustion chamber shape under lean conditions (Kuwahara et al., 1994).

Cross-Correlation PIV: With the 2-D cross-correlation function calculated from two photographs of brightness irregularity pattern in a short time interval, the velocity vectors can be determined by scanning the position of the maximum cross-correlation coefficient and knowing the time interval. The displacement vector divided by the time interval gives the velocity at one correlating area, and the smaller the size of the correlation area, the better the spatial resolution can be expected. The cross-correlation PIV results in relatively low spatial and temporal resolutions as compared with the autocorrelation PIV, but it has the advantages of an easy measurement procedure and of producing a continuous flow-field image.

Assuming that the heterogeneity of the flame luminosity would be identical with that of the flame turbulence, quantitative analyses could be performed by using high-speed direct photography and cross-correlation technique to estimate flame velocity, turbulence intensity, and scale in diesel combustion. As a result, turbulence intensity and turbulent mixing rate in high-pressure injection were larger than those of ordinary injection pressure at the early burning stage (Yamaguchi et al., 1992). The cross-correlation PIV technique was also applied to a DI diesel engine for top viewing to evaluate the gas velocities at 110 points in the combustion chamber at various swirl ratios. The data indicated that too much swirl increased the ignition delay period, and thus the amount of fuel burned in the premixed mode, resulting in an increase in NO_X emissions but a decrease in soot (Winterbone et al., 1994). To investigate the cyclic variation of the air motion in an actual engine motored at 500 rpm, the bulk flow fields inside a cup-in-piston have been quantified by the cross-correlation PIV. The ratios (R) of the root mean square (rms) to the value of ensemble-averaged velocity were used to evaluate the effect of the cyclic variation on the structure of the in-cylinder flow field. As shown in Figure 23, the values of

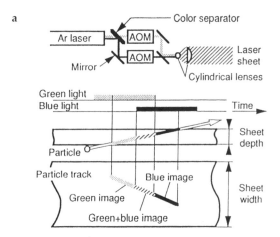

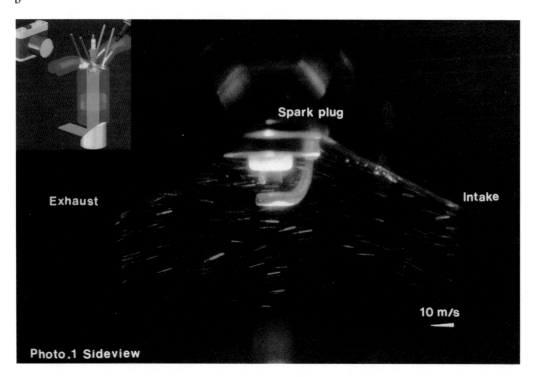

Figure 22   Two-color LLS technique for in-cylinder flow (PIV): (a) principle; (b) visualized image of in-cylinder flow.

R under the swirl condition were almost 1.0, except at the center area, and smaller than the values without swirl (Zhang et al., 1994).

Two-Color Temperature Measurement Technique

To analyze the combustion phenomena in an IC engine in more detail, the two-color technique for acquiring the 2-D images of flame temperature distribution was developed by photographing the visible rays (red and blue) radiated from the combustion products (i.e., flame or soot) and by image-processing the high speed photographs. It should be noteworthy that the 2-D images of temperature distribution thus obtained are based upon

a   b

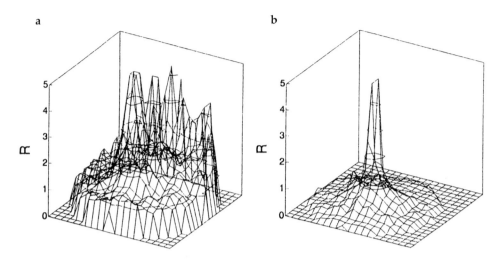

Figure 23   Comparison of R (= RMS/ensemble-averaged velocity) ratios with/without swirl in-cylinder flow at TDC and 500 rpm (PIV): (a) without swirl; (b) with swirl.

the quantity integrated along the line-of-sight of the measurement instead of the values on an LLS.

A two-color high speed shutter TV camera system was capable of recording emitted visible lights (460 nm and 640 nm) from the flame in a modified diesel engine, and then the data were put into the image processing system for conversion of the digital data. The 2-D image of the flame temperature in a gasoline engine was obtained by seeding fine particles in the intake air. The results indicated that the 2-color technique was successful for achieving 2-D quantitative images of flame temperature in both gasoline and diesel engines (Kawamura et al., 1989). The 2-D temperature distributions of flame in a disc diesel engine were also determined by the image analysis of high-speed photographs based on the above 2-color technique. As injection pressure increased, the flame temperature tended to shift to the higher value because of rapid combustion, as can be seen in Figure 24. These images of flame temperature distribution showed good agreement with the calculated ones under wide operating conditions (Kobayashi et al., 1991). To examine the effect of engine parameters, such as fuel pumping rate, hole size of injector nozzle, and injection timing on the combustion process in a DI diesel actual engine, the 2-color technique was applied to the measurements of the flame temperature. The results showed that the flame temperature was higher with both the increased pumping rate and the advanced injection timing, and that it was lower in the premixed combustion stage, but increased in the diffusion combustion stage when using a small hole area nozzle (Zhang et al., 1993).

Recently, the engine combustion phenomena were investigated by using the 2-color technique with a fiber optic cable system, which permits remote visualization of the combustion event, and is enclosed so completely that ambient light would not affect the images. The technique was easily applied to the cylinder head of a production engine, instead of the conventional optical access engines (see Figure 1 through 3). The images of flame temperature showed that the temperature first rose above 2200 K and the peak value was about 2350 K, and further, the OH radical emission intensity was a function of the flame temperature beyond about 2200 K (Shiozaki et al., 1994). Flame temperature distributions in a diesel engine were similarly visualized by the 2-color technique with a coherent fiber optic bundle of two wavelengths of 550 nm and 650 nm. The 2-D temperature images indicated that the coherent fiber-optic bundle system was also able to measure flame temperature in a production IC engine, and small-scale variations on the order of 100K were evident in the temperature images (Shakal and Martin, 1994).

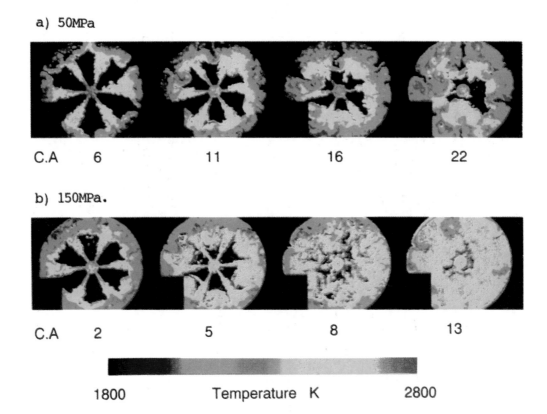

a) 50MPa

C.A   6   11   16   22

b) 150MPa.

C.A   2   5   8   13

1800   Temperature K   2800

Figure 24 2-D images of flame temperature visualized and image-processed on a diesel engine at 1000 rpm (2-color method).

Image Displaying of Measured Data

Vast quantities of flow field survey data, that is pressure, velocity, or temperature, previously would have been difficult or impossible to comprehend as it was in real time. However, they are able to be presented and interpreted by the computer-driven color video display. In the past, the data obtained in surveys of steady 3-D flow fields, e.g., wind tunnel or flight test, were first processed by such image displaying techniques as "computer graphic wake imaging system" for total pressure (Crowder, 1985) or "computer-driven color video display system" for total pressure and LDV (Winkelmann, 1989).

Engine combustion changes so rapidly that the above technique applied to the engine combustion was not found except for the following few cases. A large amount of bulk flow velocity data were obtained by using LDV measurement at 64 locations selected in the cylinder of an actual engine motored at 1000 rpm. They were image-processed to yield 2-D images of the bulk flow vectors, which generated tumble motions with various intensities by tuning the cylinder head configuration (Ando et al., 1990). When fuel spray was injected into a high pressure quiescent vessel and impinged on an inclined flat wall, vast temporal and spatial data on the droplet density of the impinging spray were measured by the laser light extinction method. Using the computed tomography (CT), these data were reconstructed to acquire 2-D images of droplet density distribution, as illustrated in Figure 25, for various inclination angles (α_w), impingement distances (z_w), and ambient pressures. The figure shows that the remarkable wall jet vortex appeared at the peripheral region of the impinging spray in the downsteam, likewise in the case of normal impingement (Saito et al., 1991).

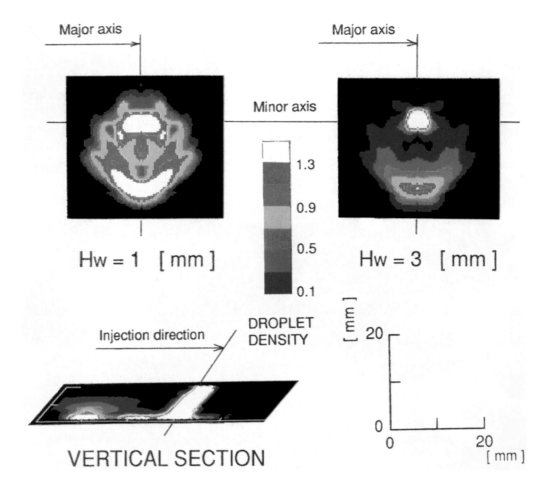

Figure 25 Applying CT method, reconstructed images of droplet density distribution in the impinging spray measured by laser light extinction method ($\alpha w = 45°$, $Zw = 24$ mm, $\rho a = 18.5$ Kg/m^3).

Image Displaying of Computational Simulation

To interpret and understand the large amounts of data calculated numerically for multi-dimensional (multi-D) simulation of engine combustion, image displaying of computed results is widely employed with vector plots, line contour plots, and wire-frame surface perspective plots. Multi-D numerical simulation techniques may be generally classified in the following categories: (1) Reynolds-averaged approaches, which are based on turbulent combustion models, (2) large eddy simulation (LES), in which small-scale perturbations are averaged, but large-scale variations are explicitly computed, and (3) direct numerical simulation (DNS), in which all scales are explicitly computed, no modeling is required since one only needs to resolve the full Navier-Stokes equations without any assumptions. In that sense, DNS appears as an exact numerical experiment. However, due to excessive computer requirements, various levels of simplification will remain necessary and hence DNS has to perform accompanying theories or experiments.

Furthermore, the image displaying of the enormous data calculated by such a computational simulation technique has recently attracted special interest owing to the following three capabilities:

1. To clarify quantitatively the transient phenomena in an internal combustion engine, which would be very difficult to visualize or impossible to reproduce.
2. To present stereographically the computed results by visual image displaying for better compromise, suitable modification, and fine regulation in the engine design concept.
3. To confirm validity of the models used in computational simulation by comparing the predicted results with the visualized photographs or computer graphics based on the measured data.

From the view of engine designers, the need for the prediction of 3-D in-cylinder flow or combustion in realistic engine geometry was growing, and further, the prediction must be made within a limited development time schedule. Although the boundary fitted coordinate is widely used for computational fluid dynamics (CFD), the mesh generation for complicated geometry is very difficult and extremely time consuming. To develop an economical CFD code for practical use, a new "Partial Cells in Cartesian coordinate" (PCC) method has been developed by considerably reducing the input data preparation time and the CPU time, because prediction is possible with fewer mesh numbers even if the objects have complicated shapes. Comparisons between the predictions and the measurements for the effective intake valve area showed good agreement at each valve lift (Takahashi et al., 1994).

SI Engine: According to the 3-D modeling of combustion in lean burn four-valve SI engines, the fluid motion in the two-intake-valve configuration is characterized by two counter-rotating vortices with inclined axes, while the one-intake-valve configuration is dominated by an inclined tumble and a high turbulence level. During the combustion phase, therefore, the one valve configuration was most favorable to the lean burn operation, and this trend was also confirmed by further experiments (Torres and Henriot, 1994). Recently, lean burn natural gas SI engines have become increasingly attractive due to their potential for low NO_x and particulate exhaust emissions and high thermal efficiency. To expedite the evolution of the combustion chamber shape in the lean burn open chamber gas engines, CFD calculation has been performed together with some experiments. Combustion progress in the modified chamber (TRI-FLOW II) was characterized by the fast growth rate of the flame kernel, and thus the flame propagated very quickly. This fast burning is also reflected in Figure 26 by the dominion of the red color almost completely filling the whole chamber at 15° ATDC, compared with the burning in a conventional chamber (Catellieri et al., 1994).

For a port-fuel-injection engine, 3-D numerical simulations of gas flow, fuel spray, and combustion have been principally conducted with the modified KIVA code. The effects of a moving valve with a stem were also adequately modeled by a novel internal obstacle treatment technique (O'Rourke and Amsden, 1987), in which the valve was represented by a group of discrete computational particles. The calculated velocity vector of intake flow through the moving valve revealed that the spray/valve interaction model was able to simulate the impingement process between the spray drops and the valve, as shown in Figure 27. Figure 27a shows the fuel spray at the end of injection and just after the intake valve opening. Afterwards, the fuel induction rate into the cylinder drastically increases and continues until BTDC as demonstrated in Figure 27b (Kuo, 1992). To estimate the atomization process of fuel droplets after the wall impingement in a port-fuel-injection engine, the Oval-Parabola Trajectories (OPT) model was developed. With OPT, the size of

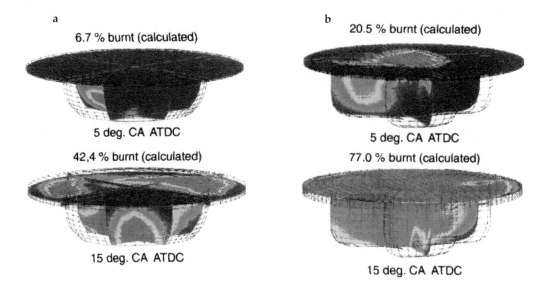

Figure 26 Comparison of combustion progress in different combustion chambers of a lean burn SI gas engine (Calculated): (a) conventional chamber; (b) TRI-FLOW II chamber.

child droplets after wall impingement and the amount of film flow remaining on the wall could be predicted in a velocity range from 2 to 40 m/s and in a region of initial diameter range below 300 μm. Figure 28 shows the details of the droplet distribution and the vapor cloud of air-fuel ratio under 14.8. A part of the injected fuel droplet breaks up into sub-droplets smaller than 10 μm due to the wall impingement. Thus the broken droplets in the cylinder move by riding on the air motion (Naitoh et al., 1994). A new spray/wall impingement model for the port-fuel-injection SI engine was developed by using the impinging spray droplets data measured with a phase Doppler particle analyzer. The model has applied to the simulation of mixture

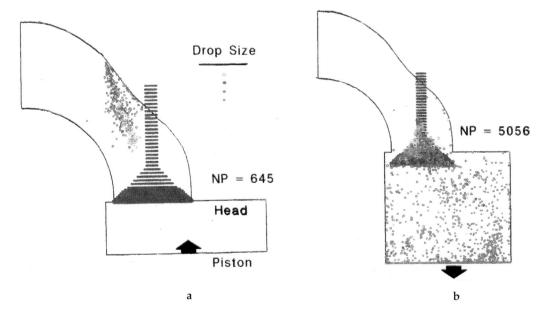

Figure 27 Front views of all computed spray parcels in port and cylinder during intake process (Calculated): (a) 50° BTDC; (b) 120° ATDC.

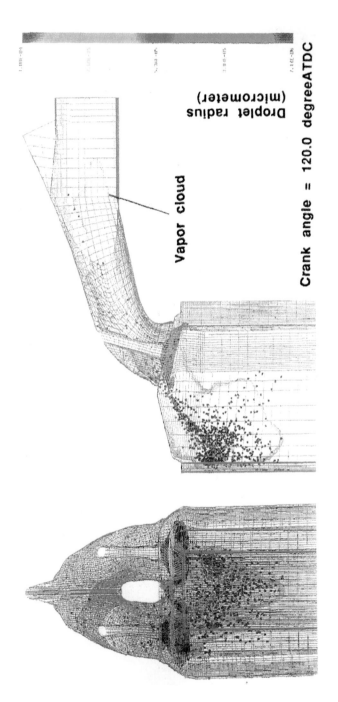

Figure 28 Image displaying for fuel droplet motions and vapor cloud during intake process in a port-fuel-injection engine (Calculated).

formation process in a lean burn gasoline engine. The results showed that the calculations of mass flow rate as well as in-cylinder velocity agreed well with the experiments, that not all the fuel injected at the intake port entered into the cylinder, and some fuels stuck on the valve and the port wall (Nagaoka et al., 1994).

To obtain the velocity vectors, fuel vapor, and gas temperature contours, 3-D computation of combustion in a premixed-charge engine has been carried out with a modified combustion submodel in KIVA code. At high delivery ratios, the results were computed using an existing combustion submodel. These results indicated a good agreement with the experiments. At low delivery ratios, however, the agreement with experiments was poor due to use of the existing submodel. To improve the disagreement, a modified combustion submodel with the same set of model constants was introduced to the computations of gas temperature contours, resulting in considerable improvements for all cases (Kuo and Reitz, 1992). On the other side, multi-D cycle simulation of flow field, combustion, and heat transfer in a SI engine was performed for the entire four stroke cycle. If the cycle simulation will repeat over several cycles until convergence, it may solve the problem of specifying the initial distribution for all flow variables. Hence the cycle simulation does not require any initial conditions, which hitherto were necessary for the conventional simulation. A simple Arrhenius model for ignition, Bray-Moss model for flame propagation, wall function model, κ-ε turbulence model for heat flux, and modified model for KIVA-II code were all employed to numerically calculate velocity vector, turbulent kinetic energy, turbulent length scale, gas temperature distribution, and species density distribution in an SI engine, as shown in Figure 29. The results showed proper trends for all the flow variables, burning rate, and heat flux (Huh et al., 1992). The multi-D computational fluid dynamics code system, FIRE, was also used to investigate the flow evolution, mixture preparation, and flame propagation in a SI engine with four valves. According to the computational results, the flow and mixture distributions near the spark plug at ignition timing were predominantly characterized by the evolution of the induction flow, because a large vortex structure (tumble) formed during the intake process and then generated several areas of high turbulence intensity during the compression stroke (Tatschl et al., 1994).

To confirm validity of simulation models, the calculated results need to be compared directly with the visualized photographs or computer graphics based on the observed or measured data. Some typical examples will be given as follows: 3-D numerical simulation was carried out to predict the gas velocity and fuel concentration profiles. The computed results showed qualitatively a good agreement with the concentration distribution measured by Rayleigh scattering technique (Kakutani et al., 1993). Turbulent combustion in a SI engine has been predicted by using direct numerical simulation (DNS) along with the framework of flamelet models, in which the flame was assumed as an ensemble of small laminar flames (flamelets) stretched and curved by turbulence. The results revealed that DNS was now applicable to the exact simulation, providing detailed information on the SI combustion process by comparison with the images of the mixture obtained with the LIF technique (Baritaud, 1994). A new model of premixed turbulent combustion has been improved by use of conditional averaging on the burned and unburned gas and by detailed analysis of the flame area evolution. The predicted results were compared with the experimental data obtained in an optical square-piston engine using the LIF, Mie scattering, and PIV techniques. Consequently, the new model proved it could correctly predict the qualitative and quantitative mixture formation during the induction process in the form of the fuel concentration distributions, as shown in Figure 30

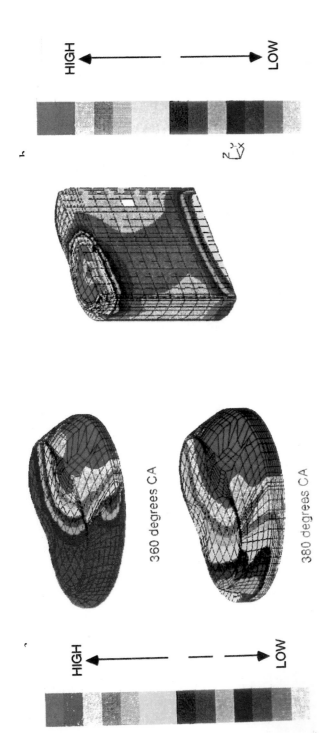

Figure 29 Multi-dimensional cycle simulation of flow and combustion in an SI engine at 1500 rpm (Calculated): (a) gas temperature distributions at various CA; (b) turbulent kinetic energy at 240° CA.

(Weller et al., 1994). Focusing on the prediction of pollutant formation, 3-D numerical computations were conducted using KIVA-II code and the standard diffusion flamelet model for three load conditions. The calculated results of gas temperature and NO-mass fraction were compared by five cuts through the combustion chamber of a DI stratified charge SI engine at 2000 rpm. The calculated results showed so correctly a correlation between NO formation and high-temperature regions that the flamelet model was found to predict the overall heat release and chemical species concentrations (or pollutant) better than the simplified combustion model (Gill et al., 1994).

Diesel Engine: The main reasons for developing a detailed mathematical model of diesel combustion are to give a better understanding of the processes involved and to have a beneficial tool for the development of a practical combustion system. The combustion process in a diesel engine, however, is more complex and less well known because of the physical phenomena involved: e.g., formation and evolution of the spray droplets, mixing of the fuel and air, auto-ignition of the gases, and propagation of the flame. Each of these phenomena requires a specific model, therefore, 3-D modeling of diesel combustion is not yet able to make predictions and the computational simulations now available are limited.

Effects of high pressure fuel injection on diesel combustion were simulated numerically by using 3-D FIRE code with several models such as turbulence, discrete droplet, and wall jet. According to the computed results, the effects of high pressure injection on the distribution of vapor and droplet in fuel spray were analyzed on a vapor mass fraction map. As shown in Figure 31, the ignition position determined on a spray temperature map also coincided well with the position visualized on an optical engine (Miwa et al., 1991). In DI diesel engines equipped with a high pressure injection system, the air-fuel mixing and the combustion of wall impinging spray were calculated by 3-D numerical simulations. As a result, design of the combustion chamber was found to control not only spray movement's direction, but also its combustion process (Endo et al., 1993). Computational simulation of DI diesel combustion has been performed through the KIVA code to examine the effect of fuel injection pressure. As fuel injection pressure increases, flame develops rapidly, and the timing of flame development near the wall becomes earlier. These experiments are predicted well by the simulation. The computed results also showed that the combustion was strongly dependent on the behavior of the fuel spray and the characteristics of the turbulence formed by the fuel spray (Sugiyama et al., 1994).

To compare with experiments for validity of the simulation models, a flamelet model coupled with an ignition delay model and the KIVA-II were used to calculate combustion in a DI diesel engine. The computed results agreed well with the experimental data, such as cylinder gas pressure and apparent heat release rate, as well as qualitative data obtained through Mie scattering, LIF, and LII techniques (Dilles et al., 1993). The effects of injection pressure and split injections on diesel engine performance and exhaust emissions (soot and NO_X) were examined by using the 3-D KIVA numerical simulation with such submodels as wave breakup atomization, drop drag, spray/wall interaction, and wall heat transfer. Consequently, spatial contours for NO_X, gas temperature, equivalence ratio, and soot in the combustion chamber at 3° ATDC were obtained, as shown in Figure 32. In the figures, these contours are superimposed for reference with the fuel droplets distribution denoted by small circles. The predicted values of cylinder pressures, heat release rates, and emission trends agreed well with the measured results over a wide range of injection characteristics (Patterson et al., 1994). Furthermore, diesel combustion in an idealized chamber was numerically simulated for examination of the effects of

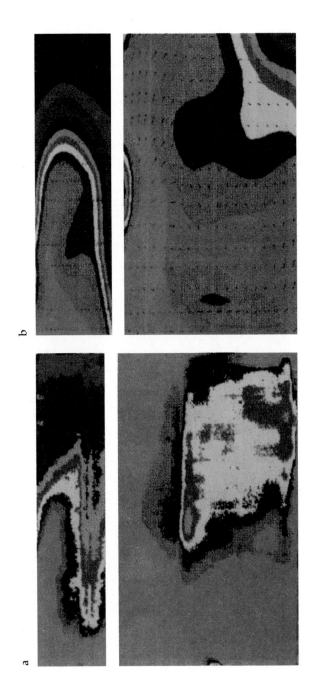

Figure 30 Comparison of ensemble-mean fuel distributions measured by LIF (a) and predicted (b) at 45 (top) and 90° (bottom) induction in a SI square engine.

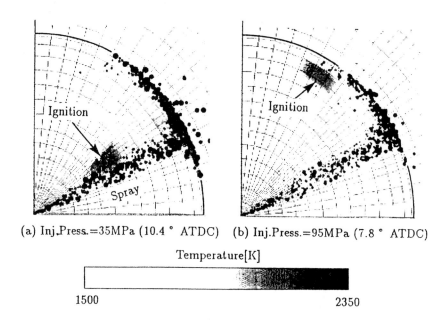

(a) Inj.Press.=35MPa (10.4 ° ATDC)   (b) Inj.Press.=95MPa (7.8 ° ATDC)

Temperature[K]

1500   2350

Figure 31   Effect of fuel injection pressure on ignition position in a diesel engine at 1800 rpm (Calculated).

several parameters, both physical and numerical, on the sensitivity of computed ignition delay. According to the results, smaller initial drop sizes resulted in shorter ignition delays and coarser meshes produced longer delays, and further, the rate of mixing had an important influence on ignition delay (Johns, 1994).

In the manner described above, the image displaying of computational simulation has been used widely to clarify the engine combustion process in more detail and to present

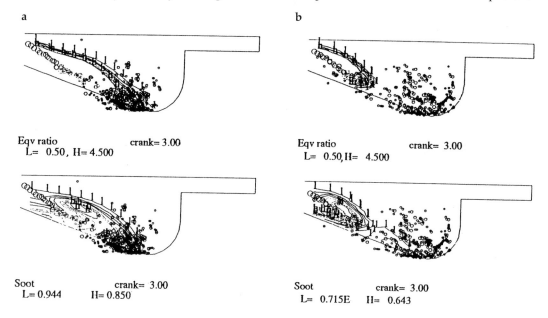

Figure 32   Comparison of spatial contours for equivalence ratio and soot between single injection (a) and split injection (b) in a diesel engine, superimposed with fuel droplet distribution (Calculated).

2-D visual images for better comprehension. One must keep in mind, however, that extensive model validation based on the single point-measurements, the conventional visualization, or the laser sheet visualizing techniques is needed before the computational simulation can be used as a predictive tool.

Conclusions for Computer Aided Visualization

Because of advanced and excellent computer resources, computer aided visualization techniques have shown a rapid rate of development and hence have become applicable to the various fields of engine combustion research as described above. Among them, the typical innovative techniques are summarized as follows:

1. PIV was first used in both motored and fired engines to obtain in-cylinder velocity vector maps, which agreed with LDV data and allowed the evaluation of vorticity and strain-rates. The PIV was even improved more by using a movie version or laser color sheet.
2. Redesigning of the intake port and combustion chamber has become very quick and easy by examination of the simulation results. The partial cells in Cartesian coordinate method was excellent for comparison with the conventional one because of considerable savings in the input data preparation time as well as the CPU time.
3. According to the simulation results of gas flow and fuel droplets in a port-fuel-injection SI engine with a moving valve, it was found that the moving valve was able to be adequately modeled by an internal obstacle treatment method.
4. Multi-D cycle simulation of flow field, combustion, and heat transfer in an SI engine can be successfully conducted for the entire 4 stroke cycles without any measured initial values, if a fast and large-scale computer system is available.
5. Although the combustion process in a diesel engine is very ambiguous and more complex, the results computed with improved models have been revealed to be reasonable for spray formation, fuel mixing, ignition, and flame propagation.

Finally, as mentioned above, there has been a drastic change in the nature of engine combustion research during the last decade. This has been caused by close cooperation between laser sheet visualization (Part I) and computer aided visualization (Part II). The former now allows high speed visualization as well as detailed image displays by aid of a computer for measuring velocities, scalar properties, species, and reacting products. The latter now allows for the development of very complex modeling approaches which offer promising prediction capabilities. Consequently, the new visualization and imaging techniques have led to a substantial improvement in our knowledge of the engine combustion process, which may reduce the time required to evaluate new design concepts in internal combustion engines, and thus, are now leading to more efficient and less polluting engines.

Abbreviations

COMODIA —	International Symposium on Diagnostics & Modeling of Combustion in IC Engine (Japan)
ICES, J	— Internal Combustion Engine Symposium, (Japan)
JSME	— Japan Society of Mechanical Engineers
SAE	— Society of Automotive Engineers, Technical Paper, (USA)
SAEJ	— Society of Automotive Engineers of Japan

References for Laser Sheet Visualization (Part I)

Alatas, B., Pinson, J. A., Litzinger, T. A., and Santavicca, D. A., 1993: A study of NO and soot evaluation in a DI diesel engine via planar imaging, *SAE, 930973.*

Ando, A. and Iida, N., 1993: Concentration measurement in a transient gas jet using laser induced fluorescence method, *11th ICES J.,* pp. 229–234.

Andresen, P., Meijer, G., Schlueter, H., Voges, H., Koch, A., Hentschel, W., Oppermann, W., and Rothe, E., 1990: Fluorescence imaging inside an internal combustion engine using tunable excimer lasers, *Appl. Opt.,* Vol. 29, No. 16, pp. 2392–2404.

Bardsley, M. E. A., Felton, P. G., and Bracco, F. V., 1988: 2-D visualization of liquid and vapor fuel in an I.C. Engine, *SAE, 880521.*

Baritaud, T. A. and Heinze, T. A., 1992: Gasoline distribution measurements with PLIF in a SI engine, *SAE, 922355.*

Baritaud, T. A., 1994: Optical and numerical diagnostics for SI engine combustion studies, *3rd COMODIA,* pp. 43–51.

Baritaud, T. A., Heinze, T. A., and Le Coz, J. F., 1994: Spray and self-iginition visualization in a DI diesel engine, *SAE, 940681.*

Bates, S. C., 1988: A transparent engine for flow and combustion visualization studies, *SAE, 880520.*

Dec, J. E., zurLoye, A. O., and Siebers, D. L., 1991: Soot distribution in a DI diesel engine using 2-D laser-induced incandescence imaging, *SAE, 910224.*

Dec, J. E. and Espey, C., 1992: Soot and fuel distributions in a DI diesel engine via 2-D imaging, *SAE, 922307.*

Eaton, A. R. and Reynolds, W. C., 1987: High-speed photography of smokee-marked flow in a motored axisymmetric engine, preprint, *Am. Soc. Mech. Eng.,* 87-FE-10.

Espey, C. and Dec, J. E., 1993: Diesel engine combustion studies in a newly designed optical-access engine using high-speed visualization and 2-D laser imaging, *SAE, 930971.*

Espey, C., Dec, J. E., Litzinger, T. A., and Santavicca, D. A., 1994: Quantitative 2-D fuel vapor concentration imaging in a firing D.I. diesel engine using planar laser-induced Rayleigh scattering, *SAE, 940682.*

Felton, P. G., Mantzaras, J., Bomse, D. S., and Woodin, R. L., 1988: Initial two-dimensional laser induced fluorescence measurements of OH radicals in an internal combustion engine, *SAE, 881633.*

Fujikawa, T., Ozasa, T., and Kozuka, K., 1988: Development of transparent cylinder engines for schlieren observation, *SAE, 881632.*

Fujimoto, M. and Tanabe, Mi., 1993: Effect of swirling flow on mixture formation in a gasoline engine (laser 2-D visualization of mixture formation) *11th ICES J.,* pp. 523–528.

Fukano, Y., Hisaki, H., Kida, S., Tachibana, K., and Kadota, T., 1993: Experimental analysis of in-cylinder flow field in a natural gas engine, *11th ICES J.,* pp. 283–288.

Gülder, O. L., Smallwood, G. J., and Snelling, D. R., 1994: Internal structure of the transient full-cone dense diesel sprays, *3rd COMODIA,* pp. 355–360.

Hoffmann, F., Bäuerle, B., Behrendt, F., and Warnatz, J., 1994: 2D-LIF investigation of hot spots in the unburnt end gas of I.C. engines using formaldehyde as tracer, *3rd COMODIA,* pp. 517–522.

Kido, A., Ogawa, H. and Miyamoto, N., 1993: Quantitative measurements and analysis of ambient gas entrainment into intermittent gas jets by laser-induced fluorescence of ambient gas (LIFA), *SAE, 930970.*

Kido, H., Nakashima, K., Kim, J., and Kataoka, M., 1993: Structure of premixed turbulent propagating flames determined by laser tomography, *Trans. JSME,* B, Vol. 59, No. 566, pp. 349–355.

Kobayashi, M., Tanabe, Y., Senda, J., and Fujimoto, H., 1993: Visualization and quantitative analysis of fuel vapor concentration in diesel spray, *11th ICES J.,* pp. 265–270.

Kosaka, H., Won, Y. H., and Kamimoto, T., 1992: A study of the structure of diesel sprays using 2-D imaging techniques, *SAE, 920107.*

Kozuka, K., Saito, A., Otsuka, M., and Kawamura, K., 1981: Television system for viewing engine combustion processes and the image analysis, *SAE, 810753.*

Lawrenz, W., Köhler, J., Meier, F., Stolz, W., Bloss, W. H., Maly, R. R., Wagner, E., and Zahn, M., 1992: Quantitative 2D LIF measurements of air/fuel ratios during the intake stroke in a transparent SI engine, *SAE, 922320.*

Lee, K., Yoo, S. C., Stuecken, T., McCarrick, D., Schock, H., Hamady, F., LaPointe, L. A., Keller, P., and Hartman, P., 1993: An experimental study of in-cylinder air flow in a 3.5 L four-valve SI engine by high speed flow visualization and two-component LDV measurement, *SAE,* 930478.

Mantzaras, J., Felton, P. G., and Bracco, F. V., 1988: Three-dimensional visualization of premixed-charge engine flames, *SAE,* 881635.

Melton, L. A., 1983: Separated fluorescence emissions for diesel fuel droplets and vapor, *Appl. Opt.,* Vol. 22. pp. 2224.

Namazian, M., Hansen, S., Lyford-Pike, E., Sanchez-Barsse, J., Heywood, J., and Rife, J., 1980: Schlieren visualization of the flow and density fields in the cylinder of a spark-ignition engine, *SAE,* 800044.

Ota, M., Hattori, H., Fujii, T., and Kadota, T., 1993: Formation and suppression of soot clouds in a D.I. diesel engine, *11th ICES J.,* pp. 31–36.

Pushka, D., Sinko, K., and Chehroudi, B., 1994: Engine-based image acquisition for piloted diesel fuel spray analysis, *SAE,* 940679.

Shimizu, R., Matsumoto, S., Furuno, S., Murayama, M., and Kojima, S., 1992: Measurement of air/fuel mixture distribution in a gasoline engine using LIEF technique, *SAE,* 922356.

Shiojim, M., Itoh, S., Yamada, O., Yamane, K., and Ikegami, M., 1992: Study of soot forming in a direct-injection diesel engine by using a laser-light sheet method, *Preprint SAE J,* 924, pp. 41–44.

Smith, J. R., 1980: Temperature and density measurements in an engine by pulsed Raman spectroscopy, *SAE,* 800137.

Steinberger, R. L., Marden, W. W., and Bracco, F. V., 1979: A pulsed-illumination, closed-circuit television system for real-time viewing of engine combustion and observed cyclic variations, *SAE,* 790093.

Tanaka, T. and Tabata, M., 1994: Planar measurements of OH radicals in an S.I. engine based on laser induced fluorescence, *SAE,* 940477.

Tomita, E., Hamamoto, Y., Tsutsumi, H., and Takasaki, S., 1993: Digital image processing of transient impinging gas jet on a flat wall and surrounding gas flow, *11th ICES J.,* pp. 535–546.

Tsue, M., Hattori, H., Saito, A., and Kadota, T., 1992: Planar fluorescence technique for visualization of a diesel spray, *SAE,* 922205.

Tsue, M., Inoue, K., Hattori, H., and Kadota, T., 1993: Studies on the structure of an unsteady gaseous jet by planar Rayleigh scattering method, *11th ICES J.,* pp. 235–249.

Winklhofer, E., Philipp, H., Fraidl, G., and Fuchs, H., 1993: Fuel and flame imaging in SI engines, *SAE,* 930871.

Wolff, D., Beushausen, V., Schlueter, H., Andresen, P., Manz, P., and Arndt, S., 1994: Quantitative 2D-mixture fraction imaging inside an I.C. engine using aceton-fluorescence, *3rd COMODIA,* pp. 445–451.

Won, Y. H., Kamimoto, T., and Kosaka, H., 1992: A study on soot formation in unsteady spray flames via 2-D soot imaging, *SAE,* 920114.

Xu, J., Behrendt, F., and Warnatz, J., 1994: 2D-LIF investigation of the early stages of flame kernel development during spark ignition, *3rd COMODIA,* pp. 69–73.

Yeh, C.-N., Kamimoto, T., Kobori, S., and Kosaka, H., 1993: 2-D imaging of fuel vapor concentration in a diesel spray via exciplex-based fluorescence technique, *SAE,* 932652.

Zhao, F. Q., Kadota, T., and Takemoto, T., 1991: Temporal and cyclic fluctuation of fuel vapor concentration in a SI engine, *SAE,* 912346.

Zhao, F. Q., Takemoto, M., Nishida, K., and Hiroyasu, H., 1993: Quantitative imaging of the fuel concentration in a SI engine with laser Rayleigh scattering, *SAE,* 932641.

Zhao, F. Q., Takemoto, M., Nishida, K., and Hiroyasu, H., 1994: PLIF measurements of the cyclic variation of mixture concentration in a SI engine, *SAE,* 940988.

Ziegler, G. W., Zettlitz, A., Meinhardt, P., Herweg, R., Maly, R., and Pfister, W., 1988: Cycle-resolved two-dimensional flame visualization in a spark-ignition engine, *SAE,* 881634.

zurLoye, A. O. and Bracco, F. V., 1987: Two-dimensional visualization of premixed-charge flame structure in an IC engine, *SAE,* 870454.

References for Computer Aided Visualization (Part II)

Ando, H., Sanbayashi, D., Kuwahara, K., and Iwachido, K., 1990: Characteristics of turbulence generated by tumble and its effect on combustion, *2nd COMODIA*, pp. 443–448.

Baritaud, T. A., 1994: Optical and numerical diagnostics for SI engine combustion studies, *3rd COMODIA*, pp. 43–51.

Cartellieri, W., Chmela, F. G., Kapus, P. E., and Tatschl, R. M., 1994: Mechanisms leading to stable and efficient combusion in lean burn gas engines, *3rd COMODIA*, pp. 17–31.

Crowder, J. P., 1985: Wake imaging system applications at the boeing aerodynamics laboratory, *SAE*, 851895.

Dillies, B., Marx, K., Dec, J., and Espey, C., 1993: Diesel engine combustion modeling using the coherent flame model in Kiva-II, *SAE*, 930074.

Endo, H., Nakagawa, H., Mori, S., and Oda, Y., 1993: Numerical study on diesel spray combustion after wall impingement, *11th ICES J.*, pp. 451–456.

Gill, A., Gutheil, E., and Warnatz, J., 1994: Numerical investigation of the turbulent combustion in a direct-injection stratified charge engine with emphasis on pollutant formation, *3rd COMODIA*, pp. 583–588.

Hentschel, W. and Stoffregen, B., 1987: Flow visualization with laser light-sheet techniques in automotive research, *4th Flow Visualization*, Véret, C., ed., Hemisphere, pp. 839–846.

Huh, K. Y., Kim, K. K., and Min, K., 1992: Multidimensional cycle simulation of flow field, combustion and heat transfer in a spark ignition engine, *SAE*, 920588.

Johns, R. R., 1994: Calculation of diesel combustion in idealised chambers, *3rd COMODIA*, pp. 299–305.

Kakutani, H., Tsue, M., Idokawa, M., and Kadota, T., 1993: Studies on the mixture formation process in a gas engine, *11th ICES J.*, pp. 609–614.

Kawamura, K., Saito, A., Yaegashi, T., and Iwashita, Y., 1989: Measurement of flame temperature distribution in engines by using a two-color high speed shutter TV camera system, *SAE*, 890320.

Kobayashi, S., Sakai, T., Nakahira, T., and Tujimura, K., 1991: Measurement of flame temperature distribution in DI diesel engine with high pressure fuel injection, *9th ICES J.*, pp. 115–126.

Kuo, T.-W., 1992: Multidimensional port-and-cylinder gas flow, fuel spray, and combustion calculations for port-fuel-injection engine, *SAE*, 920515.

Kuo, T.-W. and Reitz, F. D., 1992: Three-dimensional computations of combustion in premixed-charge and direct-injected two-stroke engines, *SAE*, 920425.

Kuwahara, K., Kawai, T., and Ando, H., 1994: Influence of flow field structure after the distortion of tumble on lean-burn flame structure, *3rd COMODIA*, pp. 89–94.

Lee, J. and Farrell, P. V., 1993: Intake valve flow measurements of an IC engine using particle image velocimetry, *SAE*, 930480.

Miwa, K., Nakakita, K., Ohsawa, K., and Watanabe, S., 1991: 3-D numerical study of mixture-formation and ignition processes in a DI diesel engine with high pressure fuel injection, *9th ICES J.*, pp. 127–131.

Nagaoka, M., Kawazoe, H., and Nomura, N., 1994: Modeling fuel spray impingement on a hot wall for gasoline engines, *SAE*, 940525.

Naitoh, K., Takagi, Y., Kokita, H., and Kuwahara, K., 1994: Numerical prediction of fuel secondary atomization behavior in SI engine based on the oval-parabola trajectories (OPT) model, *SAE*, 940526.

Nino, E., Gajdeczko, B. F., and Felton, P. G., 1992: Two-color particle image velocimetry applied to a single cylinder two-stroke engine, *SAE*, 922309.

Nino, E., Gajdeczko, B. F., and Felton, P. G., 1993: Two-color particle image velocimetry in an engine with combustion, *SAE*, 930872.

O'Rourke, P. J. and Amsden, A. A., 1987: Three dimensional numerical simulations of the UPS-292 stratified charge engine, *SAE*, 870597.

Patterson, M. A., Kong, S.-C., Hampson, G. J., and Reitz, R. D., 1994: Modeling the effects of fuel injection characteristics on diesel engine soot and NO_x emissions, *SAE*, 940523.

Reeves, M., Garner, C. P., Dent, J. C., and Halliwell, N. A., 1994: Particle image velocimetry measurements of barrel swirl in a production geometry optical IC engine, *SAE*, 940281.

Reuss, D. L., Adrian, R. J., Landreth, C. C., French, D. T., and Fansler, T. D., 1989: Instantaneous planar measurements of velocity and large-scale vorticity and strain rate in an engine using particle-image velocimetry, *SAE*, 890616.

Reuss, D. L., Bardsley, M., Felton, P. G., Landreth, C. C., and Adrian, R. J., 1990: Velocity, vorticity and strain-rate ahead of a flame measured in an engine using particle-image velocimetry, *SAE*, 900053.

Saito, M., Hashimoto, A., Nagae, M., Senda, J., and Fujimoto, H., 1991: Characteristics of diesel spray impinging on a flat wall, (3rd Report), *Trans. JSME*, (B), Vol. 57, No. 543, pp. 333–340.

Shakal, J. S. and Martin, J. K., 1994: Imaging and spatially resolved two-color temperature measurements through a coherent fiberoptic: observation of auxiliary fuel injection effects on combustion in a two-stroke DI diesel, *SAE*, 940903.

Shiozaki, T., Miyashita, A., Aoyagi, Y., and Joko, I., 1994: The analysis of combustion flame in a DI diesel engine (part 2 - hydroxyl radical emission versus temperature), *3rd COMODIA*, pp. 523–528.

Stolz, W., Köhler, J., Lawrenz, W., Meier, F., Bloss, W. H., Maly, R. R., Herweg, R., and Zahn, M., 1992: Cycle resolved flow field measurements using a PIV movie technique in a SI engine, *SAE*, 922354.

Sugiyama, G., Ryu, H., and Kobayashi, S., 1994: Computational simulation of diesel combustion with high pressure fuel injection, *3rd COMODIA*, pp. 391–396.

Sweetland, P. and Reitz, R. D., 1994: Particle image velocimetry measurements in the piston bowl of a DI diesel engine, *SAE*, 940283.

Takahashi, Y., Fukuzawa, K., and Fujii, I., 1994: Numerical simulation of flow in intake ports and cylinder of multi-valve S.I. engine using PCC method, *3rd COMODIA*, pp. 529–534.

Tatschl, R., Wieser, K., and Reitbauer, R., 1994: Multidimensional simulation of flow evolution, mixture preparation and combustion in a 4-valve SI engine, *3rd COMODIA*, pp. 139–149.

Torres, A. and Henriot, S., 1994: 3D modeling of combustion in lean burn four-valve engines: influence of intake configuration, *3rd COMODIA*, pp. 151–156.

Weller, H. G., Uslu, S., Gosman, A. D., Maly, R. R., Herweg, R., and Heel, B., 1994: Prediction of combustion in homogeneous-charge spark-ignition engines, *3rd COMODIA*, pp. 163–169.

Winkelmann, A. E., 1989: Flow-field survey data, *Handbook of Flow Visualization*, Yang, W. J., ed., Hemisphere, pp. 363–374.

Winterbone, D. E., Yates, D. A., Clough, E., Rao, K. K., Gomes, P., and Sun, J. H., 1994: Quantitative analysis of combustion in high-speed direct injection diesel engines, *3rd COMODIA*, pp. 261–267.

Yamaguchi, I., Shioji, M., and Tsujimura, K., 1992: An analysis of flow and turbulence in the combustion field for D.I. diesel engine with high pressure fuel injection with an image processing technique, *Trans. SAEJ*, Vol. 23, No. 2, pp. 109–114.

Zhang, L., Minami, T., Takatsuki, T., and Yokota, K., 1993: An analysis of the combustion of a DI diesel engine by photograph processing, *SAE*, 930594.

Zhang, L., Ueda, T., Takatsuki, T., and Yokota, K., 1994: A study of the cycle-to-cycle variation of in-cylinder flow on a motored engine, *3rd COMODIA*, pp. 541–546.

chapter two

Visualization of the Flow Into and Out of a Hole in a Duct Wall

L. A. S. B. Martins and J. H. Whitelaw

Imperial College of Science, Technology and Medicine, U.K.

Abstract — The influence of the ratio of the velocity of air flowing through a hole in a duct wall to that in the duct, is quantified in terms of visualization of the surface of the duct and of the resulting jets. The velocity ratio was modified by an obstruction in the downstream region of the duct, as in a gas-turbine combustor, and a high velocity ratio corresponding to a large blockage gave rise to a substantial region of reverse flow with the resulting jet emerging normal to the plate as would be expected from the near-plenum conditions of the duct. Velocity ratios close to unity, with lower blockage in the duct, caused a region of separated flow immediately downstream of the hole with jets which were angled in the direction of the duct flow.

Introduction

The flow into the combustor of gas-turbine engines is distributed and directed by the shape of the diffuser and the shape of the duct which surrounds the combustor. It is important, for example, that the jets which supply the primary region impinge with a substantial proportion of the mass flow directed into the primary vortex to mix with air from the swirler and fuel from the centrally located spray. Similarly, the jets to the secondary region are required to mix with the hot products with a trajectory which is in the general direction of the flow. Thus, it is important to know the relationship between the trajectory of the jets and the duct geometry. Also, information is required about the nature of the flow around the hole and within the duct mainly to understand the reasons for the trajectory and discharge coefficient and also to guide the development of calculation methods.

Many experiments have been performed to determine discharge coefficients. The early works include those of Dittrich and Graves (1956) and of Kaddah (1964) which resulted in correlation formulae. It is evident, however, that the correlations do not account for all parameters and, as a consequence, the experimental scatter is large.

The more recent correlation of Adkins and Gueroui (1986) clarifies some of the parameters which can be important. They concluded that the discharge coefficient and angle of the jet to that of the hole are a function of the ratio of the maximum jet velocity, V_j, to the average velocity in the duct, U_d, provided the static pressure ratio is less than 1.06, the ratio of wall thickness to hole diameter is less than 0.1, the hole Reynolds number is greater than 10^5, and the height of the duct is sufficient to ensure that pressure gradients are small. It can be expected that the requirement for the hole Reynolds number is high except where the characteristic dimension of the duct is small compared to that of the hole.

The use of calculation methods, based on the numerical solution of conservation equations in differential form, to represent combustor flows has been reported by Palma and McGuirk (1992) and Chow and McGuirk (1991). In the latter paper, which considered the flow in the small combustor investigated experimentally by Chow and Whitelaw (1992), the trajectory of the dilution jets was determined by comparing calculated trajectories with streak-line visualization and found to be some 10° different from that of the axis of the holes. No attempt was made to calculate the flow in the annulus, but this is likely in the near future and supporting measurements will be required.

The results presented here were obtained by visualization of the near-wall flow around a hole, as represented by surface streaks, and of the jet emerging from the hole which was located in the wall of a duct comprising two parallel sides and a downstream blockage to allow velocity ratios from 1.3 to greater than 35. The range of flow conditions shown in Table 1 is similar to that considered by Graves and Grobman (1958) who calculated velocity ratios of about 1.2 and infinity for the primary and dilution jets of a tubular combustor, and is representative of those found in the combustors of many gas-turbine engines. The following section describes the flow configuration and measurement techniques and the third section presents the results. The paper ends with a summary of the more important findings.

Flow Configuration and Visualization Method

The small purpose-built wind tunnel comprised a centrifugal blower driven by a variable-speed DC motor, a wide-angle diffuser expanding in one plane, a settling chamber with honeycomb and two screens, and a two-dimensional contraction leading to the test section (Figure 1) which was fabricated from 12.5 mm-thick perspex. The contraction had an aspect ratio of five and conformed to the recommendations of Bradshaw (1972). The test section was 800 mm long, with a constant cross-section of 360 by 45 mm and a 30 mm diameter hole was located in the center of the bottom wall, 500 mm from the end of the contraction. The edges of the hole were sharp and free from burrs. A valve was located at the end of the test section and comprised two movable aluminum sleeves of 4 mm thickness which could be adjusted to provide symmetrically located exit apertures from 0 to 45 mm.

The hole Reynolds number varied from 3.4×10^4 to 1.2×10^5 depending on the velocity ratio as shown in Table 1, and ensured that the jets were turbulent. The ratio of wall thickness to hole diameter was 0.42, the static pressure ratio was always less than 1.04, and the ratio of duct height to hole diameter 1.5, so that it might be expected that the velocity ratio would be the only important parameter.

The mass flow in the duct was determined from the static pressure drop in the contraction which was calibrated in terms of integrated velocity profiles measured with a Pitot tube in its exit plane. This procedure provided average velocities which were accurate

Table 1   Flow Conditions Investigated

Fraction of valve open	V_j/U_d	$Re_d \times 10^4$
Closed	35	12
1/9	9.3	11
1/3	4.1	9.1
2/3	2.0	5.8
8/9	1.3	3.4

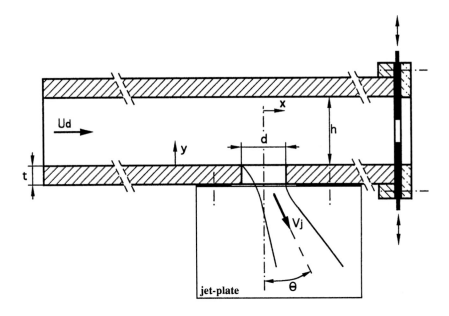

Figure 1 Arrangement of the perspex duct.

to better than 4%, except at the highest velocity for which the flow rate and pressure drop were very small and the uncertainty was of the order of 10%. The profile of the jet velocity was measured in its exit plane, also with a Pitot tube, giving an accuracy of better than 1% for the maximum jet velocity. All pressures were measured with a Betz projection micrometer with an accuracy of 0.1 mm of water, corresponding to a maximum uncertainty of 1.4 and 0.1% in the duct and maximum jet velocities respectively.

The surface flow was visualized with a thin layer of an emulsion of titanium dioxide, kerosene, oleic acid, and silicone oil which was applied evenly on the surface of the plate and in the vicinity of the hole. The air was arranged to flow and caused the emulsion to streak according to the flow direction and to collect in regions of stagnation. The flow patterns were observed through the upper wall and, when established, photographs were taken. The results are qualitative and should be interpreted with care and only as representative of the flow close to the wall. Nevertheless, trajectories and stagnation regions are readily established as will be shown.

The jets emerging from the hole were visualized in the vertical plane delineated by the geometric axes of the duct and of the hole (xy plane, z = 0, see Figure 1). For this purpose, a 0.8 mm thick steel plate was firmly bolted to the outside part of the duct lower wall and on the longitudinal axis of the hole. The same emulsion paint as described above was used and similar procedures were followed. Although some gravity effects are possible, the jet visualization results clearly show the potential core regions and allow for the evaluation of the jet deflection angles as discussed in the next section.

Results

Photographs of the surface flow on the lower plate and around the hole are presented in Figure 2 for the flow from left to right, and of the resulting jets in Figure 3 for the velocity ratios from 1.3 to 35.

With the highest velocity ratio, for which the slit valve was closed, the flow entered as if from a plenum chamber and there is a clear tendency for rotation of the flow as suggested by Lefebvre (1983) for cases where the pitch of the holes is larger than the duct height, as

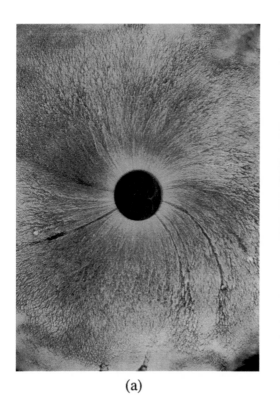

(a)

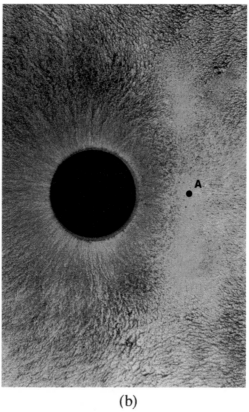

(b)

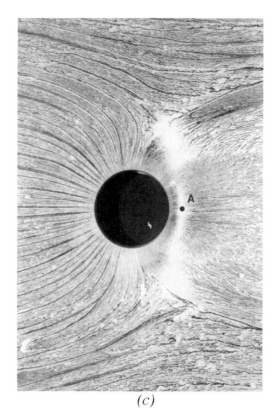

(c)

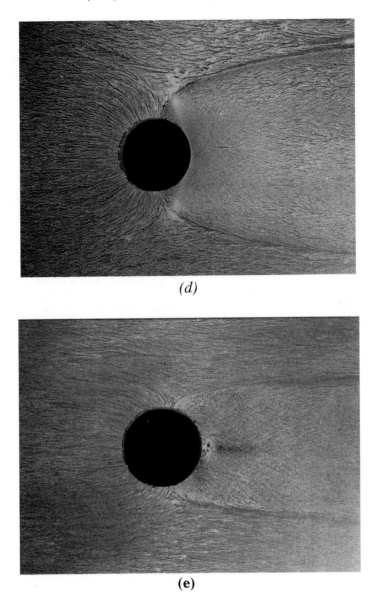

(d)

(e)

Figure 2   Surface visualization on the duct lower wall: (a) $V_j/U_d = 35$; (b) $V_j/U_d = 9.3$; (c) $V_j/U_d = 4.1$; (d) $V_j/U_d = 2.0$; (e) $V_j/U_d = 1.3$.

in this case. The corresponding jet of Figure 3a is symmetric with its axis coincident with that of the hole.

There is no vortex in the photograph, Figure 2b, which corresponds to a velocity ratio of 9.3 and to an opening of the slit valve of just over 0.1 of the duct height. A broad and curved impingement region exists downstream of the hole with a stagnation point in the plane of symmetry and nearly one diameter downstream of the edge of the hole (Point A). The flow into the hole comes from all directions. The corresponding jet of Figure 3b has an axis which is curved and initially at an angle of around 7° to the axis of the hole.

Reduction of the velocity ratio to 4.1, Figures 2c and 3c, with the slit valve open to around 0.3 of the duct height, has caused the curved attachment line downstream of the hole to move closer to the hole so that the flow into the hole is clearly less radially

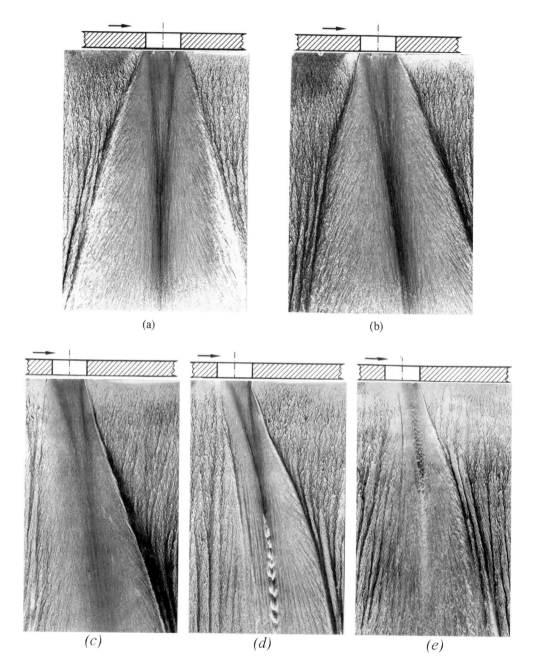

Figure 3 Visualization of the jet longitudinal plane: (a) $V_j/U_d = 35$; (b) $V_j/U_d = 9.3$; (c) $V_j/U_d = 4.1$; (d) $V_j/U_d = 2.0$; (e) $V_j/U_d = 1.3$.

symmetric. The stagnation point is now only 5 mm from the edge of the hole and the attachment region is thinner. The resulting jet has an angle of more than 10° to the axis of the hole and the initial region of the potential core has been moved in the downstream direction.

With velocity a ratio of 2.0, Figures 2d and 3d, the jet was deflected by around 14°, with the initial region of the potential core moved well downstream of the edge of the hole to provide a thinner initial jet and curvature of the edge of the jet particularly obvious as it

Table 2 Jet Deflection Angle θ. Comparison
between Observed and Theoretical Values

V_j/U_d	Observed angle θ (°)	Arcsin (U_d/V_j) (°)
35	0.0–2.5	1.6
9.3	7.0	6.2
4.1	10.5	14.1
2.0	13.5	30.0
1.3	8.5	60.3

moves towards a more symmetrical arrangement far downstream. The jet results suggest that there is a substantial region of separated flow close to the upstream edge of the hole; this is also clear on Figure 3e and the two jet trajectories suggest a tendency for the jet angle to reduce with further reduction in velocity ratio. The surface visualization shows that the stagnation point is very close to the downstream edge of the hole with the attachment line curving upstream of this point. With the velocity ratio of 1.3, the low flow rate is evident in the less defined streaklines and there is no evidence of attachment or of reverse flow apart from two small separation bubbles at the downstream edge of the hole and on either side of the line of symmetry; these separated-flow regions were also found at lower velocities.

Table 2 lists the angles of the jets and those calculated from the formula of Adkins (1986) and it is evident that the measurements deviate from the correlation at the lower velocity ratios. This may be associated to the likely regions of separated flow with the low velocity ratios and they, in turn, are likely to be related to the thickness of the plate. Dittrich and Graves (1956) found that an increase in the thickness of the plate led to a decrease in the discharge coefficient for velocity ratios smaller than 4, and this may also stem from the separation.

Concluding Remarks

The large jet velocities gave rise to a substantial region of reverse flow in the duct with the resulting jet emerging normal to the plate, as would be expected from the near-plenum conditions of the duct. With no bulk flow in the duct downstream of the hole, the surface visualization indicated a swirling motion. Low velocity ratios led to a region of separated flow immediately downstream of the hole with jets which were angled in the direction of the duct flow; the separated flow in and downstream of the hole was probably a feature of the hole geometry with is comparatively high ratio of plate thickness to hole diameter. For these cases, the surface visualization indicated the disappearance of the region of attachment and the formation of small vortices. It is evident that the velocity ratio is the main parameter by which to describe the discharge coefficient and jet trajectory at high velocity ratios and that, at lower ratios, the separated flow, probably induced in part by the thickness of the duct wall, becomes increasingly important. It is likely that the *a priori* calculation of jets will require knowledge of the details of the hole and the thickness of the wall as well as the ability to represent the attachment, reverse flows, and vortices observed here in terms of surface streaks.

Acknowledgments

The authors are indebted to the Junta Nacional de Investigação Científica e Tecnológica and to Rolls Royce Plc for their financial support. They would also like to thank the department technical staff of Imperial College, in particular Mr. P. M. Jobson and Mr. P. Trowell for their help and assistance.

References

Adkins, R. C. and Gueroui, D., 1986: An improved method for accurate prediction of mass flows through combustor liner holes, *J. Eng. Gas Turbines Power*, Vol. 108, pp. 491–497.

Bradshaw, P., 1972: Two more wind tunnels driven by aerofoil-type centrifugal blowers, *Imperial College Aero Report*, 72–10.

Chow, S. K. and McGuirk, J. J., 1991: Numerical prediction of flow and combustion characteristics of a model annular combustor, Presented at the 36th ASME International Gas Turbine and Aeroengine Congress and Exposition, Orlando.

Chow, S. K. and Whitelaw, J. H., 1992: Scalar characteristics in a liquid fuel combustor with a curved exit nozzle, in *Aerothermodynamics in Combustors*, Lee, R. S. L., ed., Springer-Verlag, pp. 291–300.

Dittrich, R. T. and Graves, C. C., 1956: Discharge coefficients for combustor-liner air-entry holes. I-Circular holes with parallel flow, *NACA*, TN 3663.

Graves, C. C. and Grobman, J. S., 1958: Theoretical analysis of total-pressure loss and airflow distribution for tubular turbojet combustors with constant annulus and liner cross-sectional areas, *NACA Report*, 1373.

Lefebvre, A. H., 1983: *Gas Turbine Combustion*, 1st Ed., McGraw-Hill.

Kaddah, K. S., 1964: Discharge coefficients and jet deflection studies for combustor-liner air-entry holes, *MSc Thesis*, Cranfield Institute of Technology.

Palma, J. M. L. M. and McGuirk, J. J., 1992: The influence of numerical parameters on the calculation of gas turbine combustor flows, *Comput. Meth. Appl. Mech. Eng.*, Vol. 96, pp. 65–92.

chapter three

An Experimental Study on Transition and Mixing Processes in a Coaxial Jet

H. Yamashita, G. Kushida, and T. Takeno

Department of Mechanical Engineering, Nagoya University, Chikusa-ku, Nagoya, Japan

Abstract — An experimental study was made on the transition and mixing processes of a coaxial water jet into a coflowing water stream. The jet was visualized by means of laser induced fluorescent technique, and time dependent behavior of the jet was studied to understand the transition mechanism from laminar to turbulent flow, and the mixing characteristics in the downstream transitional and turbulent regions. The transition was found to be caused by a helical mode Kelvin-Helmholtz instability. The observed distributions of concentration probability density function and of scalar dissipation rate suggest that the mixing process in the transitional region is governed mostly by the large scale flow fluctuation induced by the instability, but the role of small scale turbulence becomes more and more important in the downstream turbulent region.

Introduction

In the past a great number of experimental studies were conducted to understand the transition from laminar to turbulent flow and turbulent mixing processes in coaxial jets. However, recent remarkable progresses in flow visualization techniques by use of lasers have renewed our attention to this rather old problem. With the aid of these novel techniques, and of the associated data processing techniques, there is a fair chance that we can obtain a more profound understanding of these phenomena; e.g. Dahm et al. (1991). The objective of the present study is to apply the planar laser-induced fluorescence technique to visualize the time dependent behavior of a coaxial jet to understand the transition and mixing processes in the jet.

Experimental

The experiment was conducted by using the setup shown schematically in Figure 1. The horizontal water tunnel had a square cross-section of 100 mm × 100 mm with length of 1000 mm, and the wall was made of acrylic resin and was transparent to allow for laser access and to collect induced fluorescence. In the axis of the tunnel an injector for water injection with a fluorescent dye (disodium fluorescein) was placed coaxially. The injector was a straight tube made of stainless steel and was 800 mm in length, 6.0 mm and 7.0 mm in inside and outside diameter, respectively. The injector tip was configured to make a sharp

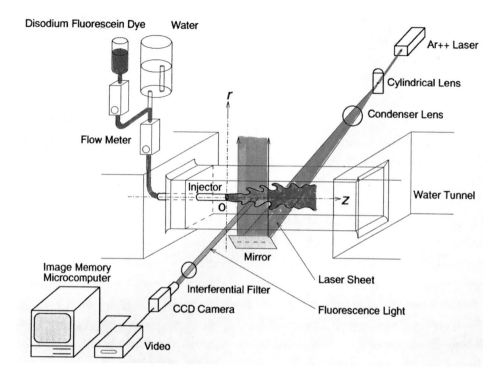

Figure 1   Schematic of experimental setup.

edge to avoid formation of a vortex, as is shown in Figure 2. The dye in the jet was excited by the irradiation of 488.0 nm emission from 1 W Ar++ laser. The optics for the two-dimensional laser sheet consisted simply of a cylindrical lens and a condenser lens. A laser beam of 1.4 mm diameter was expanded by the cylindrical lens to make a sheet and the thickness of the sheet was made thinner by the condenser lens to give the maximum value of 0.50 mm across the visual field of the recording camera. The sheet was passed through the jet axis from the bottom of the tunnel through a plane mirror, and the induced fluorescence was collected at a right angle to the sheet and was recorded by a CCD camera. In most experiments, the shutter speed of the camera was 1/1000 s and some experiments were made with a speed of 1/10,000 s to check effects of the exposure time. The smearing effects for 1/1000 s were not serious, so then the time resolution was considered good enough to obtain instantaneous concentration fields. An interference filter was placed in front of the camera to block the original laser light scattered by small particles inevitably

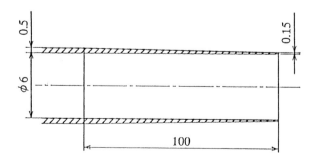

Figure 2   Detailed configuration of injector tip.

contained in the flow. The fluorescence intensity was digitized through an image memory and was analyzed in a microcomputer to yield the instantaneous concentration field.

The experiment was conducted for two different velocities of coflowing water: $U = 0.02$ and 0.04 m/s. The injection velocity u_o was varied from 0.02 to 0.80 m/s. The corresponding injection Reynolds number $Re = u_o d/v$, where d and v are inside diameter of the injector and kinematic viscosity of the jet fluid respectively, was from 120 to 4800. The diffusivity of the dye into water is very small, therefore the value of Schmidt number is around 2000. Then the limiting scale in the turbulent concentration field, the so-called Batchelor scale (Batchelor, 1958), is in the order of 0.001 mm. This gives rise to the serious spatial resolution problem of the present visualization technique. One pixel size of a CCD camera corresponded to 0.17 mm in actual size. Moreover, the thickness of the laser sheet was 0.50 mm and these sizes are much larger than the limiting scale of a fully developed turbulent flow. The relation between the laser thickness and the representative scale of the concentration field is shown schematically in Figure 3 for two cases. The left one shows the case when the representative scale is larger than the thickness, and hence the pixel size, and the concentration field can be resolved spatially by the present method. The right one, on the other hand, shows the case when the representative scale is smaller than the thickness. In this case, the signal from respective pixels represents just the concentration averaged over the laser thickness. Therefore, the visualized jet represents the correct concentration field in the upstream laminar and transitional regions, where the representative scale should be larger than 0.50 mm. In the further downstream turbulent region, however, the visualized jet must be accompanied by some averaging effects since the three-dimensional small scale turbulence can be as small as the Batchelor scale. Then we have to be very careful in the interpretation of the obtained concentration field. This inevitably makes the present visualization technique lose some important information, yet it still can give us much valuable information about instantaneous concentration field.

The digitized signal fed to the microcomputer, as it is, does not necessarily give the correct concentration. We have to correct the signal for the following three factors; sensitivity characteristics of CCD camera, intensity distribution of laser light source, and self-adsorption of laser light. In the present experiment, it was found that the last effect could be avoided by decreasing the dye concentration in the jet. However, the remaining two effects had to be corrected. The first factor was corrected by making a calibration experiment with a cell of known uniform concentration. The response of the CCD camera, including those of the video recorder and the image memory, to a change in the intensity of light source was studied to give a sensitivity characteristic curve to make the correction

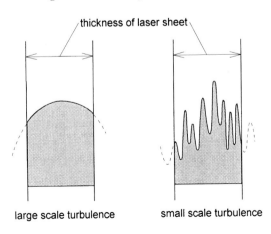

large scale turbulence   small scale turbulence

Figure 3   Schematic representation of relationship between laser thickness and turbulence scale.

of the first factor. The second factor came from a Gaussian-like intensity distribution of the laser source. The tunnel was filled with water of uniform concentration to the dye, and the signal from each pixel in the visual field of CCD camera was studied to determine the correction factor for that pixel.

Figure 4 shows how these factors affect the obtained planar concentration field. The color scale represents the relative concentration of the dye normalized by that of the jet fluid, while axial distance Z from the injector exit and radial distance r from the axis are made nondimensional by inside diameter d of the injector. Figure 4 (a) shows the uncorrected concentration field based on the original signal from the image memory. Figure 4 (b) shows the concentration field after the correction for the first effect alone, while (c) shows that after the correction for the two effects. The comparison between (a) and (c) will reveal the significance of these factors.

Results and Discussions

Effects of Reynolds Number

Figure 5 shows the dependence of the instantaneous concentration field and of the corresponding time averaged concentration field on the Reynolds number. The time average was taken for 256 instants. The coflowing water velocity was kept at $U = 0.04$ m/s and the injection velocity was changed. When Re is small, the whole flow field remains laminar and steady. The small diffusivity of the dye prevents the lateral expansion of the jet fluid by molecular diffusion, and the jet remains almost straight in the axial direction. When Re is increased, a sinusoidal fluctuation appears along the jet and the flow becomes turbulent in the downstream. This is the transition from laminar to turbulent flow, and we may define the transition point by a method described later. The transition point moves upstream with a further increase in Re. As is seen in the figure, the mixing in the transitional region, just downstream of the transition point, appears to be governed mostly by the large scale fluctuations developed from the sinusoidal fluctuation, whereas in the further downstream the contribution of small scale turbulence appears to be significant. As can be seen in the figure, the time-averaged concentration fields are always axisymmetric.

Transition

In order to understand the transition mechanism, time dependent behavior of the jet was studied. The behavior for $Re = 1500$ is shown in Figure 6. The time between each sequence is $1/30$ s. As is seen in the figure, the point where the sinusoidal fluctuation first appears moves back and forth. This behavior was found to be closely related to the pipe flow turbulence, which was induced inside the injector and was ejected into the jet. To define the transition point from laminar to turbulent flow, a probability density function (PDF) was introduced. It represents the number of times when the concentration on the axis first decreases by certain percentage from that of the jet fluid. Figure 7 shows how the PDF depends on axial distance from the injector exit for the case when the decrease was 20%. The PDF was derived from 4096 instants data for respective values of Re. When Re is smaller than 2250, the pipe flow inside the injector remains laminar and the PDF exhibits a Gausian-like distribution with a single peak. As Re is increased, the peak location shifts upstream, indicating that the transition tends to be caused in the upstream. The distribution becomes bimodal when Re is 2400, for which the pipe flow is in the transition regime from laminar to turbulent flow, and the turbulence is ejected into the jet intermittently. For further increases in Re, the distribution takes a Gausian-like distribution again and the

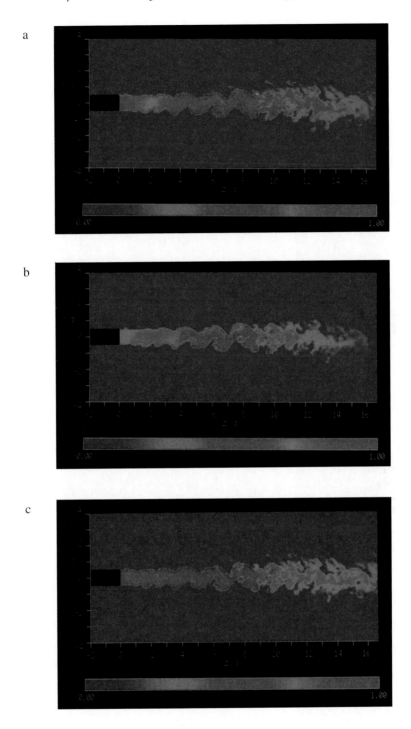

Figure 4 Corrections for sensitivity characteristics of CCD camera, intensity distribution of laser light source: (a) original output of image memory; (b) corrected for sensitivity characteristics of CCD camera; (c) corrected for sensitivity characteristics of CCD camera and intensity distribution of light source.

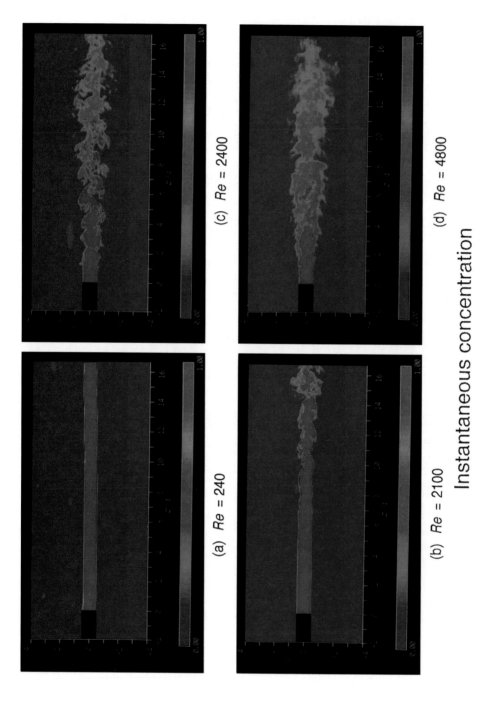

(a) *Re* = 240

(b) *Re* = 2100

(c) *Re* = 2400

(d) *Re* = 4800

Instantaneous concentration

Figure 5   Dependence of instantaneous and time averaged concentration fields on Reynolds number.

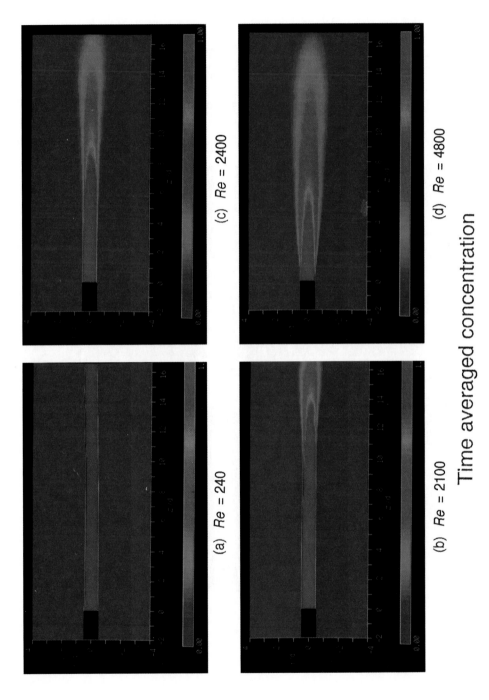

(a) *Re = 240*

(b) *Re = 2100*

(c) *Re = 2400*

(d) *Re = 4800*

Time averaged concentration

Figure 5 (continued)

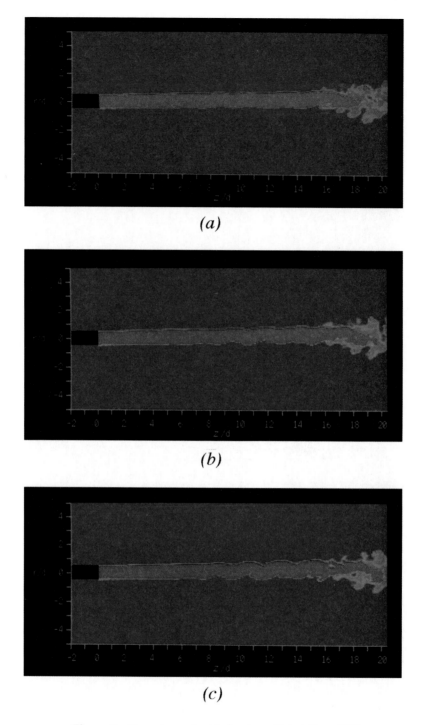

(a)

(b)

(c)

Figure 6   Time-dependent behavior of jet at $Re = 1500$.

peak location moves upstream slowly approaching an asymptotic value around $4d$. We may define the transition length Z_t by the arithmetic average of the PDF distribution, and its dependence on Re is shown in Figure 8. As is seen in the figure, Z_t decreases with Re in the laminar and transition regimes of pipe flow, and becomes almost constant in the fully developed turbulent region.

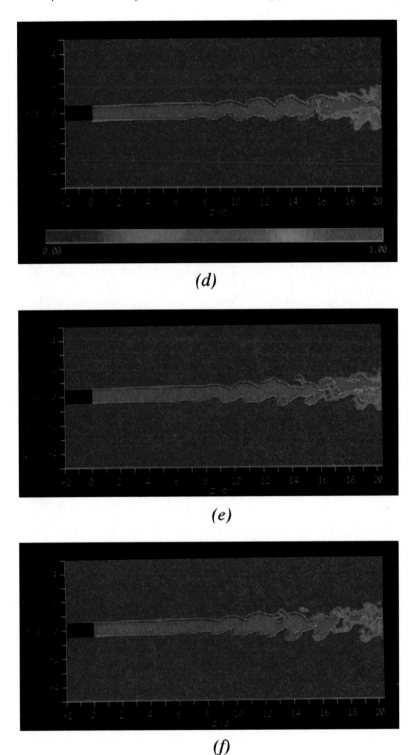

(d)

(e)

(f)

Figure 6 (Continued)

One important observation related to the transition is that the fluctuation appeared asa sinusoidal form. This was very clear when *Re* was relatively small and the transitionbehavior was easy to observe. Since we are looking at the jet behavior in the

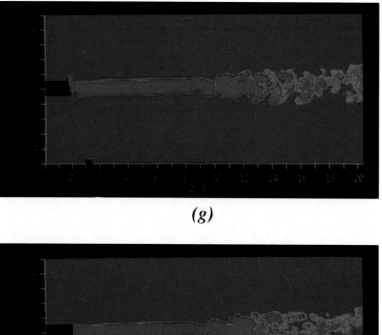

(g)

(h)

Figure 6 (Continued)

two-dimensional plane, cut by a laser sheet passing through the axis, this means that the induced three-dimensional fluctuation is spiral or helical. That is, the helical mode is most unstable in the Kelvin-Helmholtz instability in the jet being studied. This observation is consistent with the prediction of our recent numerical study (Kushida et al., 1993).

Mixing Characteristics

The mixing characteristics in the transitional and turbulent portions of the jet were studied by means of the concentration PDF. Figure 9 shows the PDF distributions in axial distance-concentration plane for $Re = 2700$ and 4800. The corresponding distributions for the fixed axial distance are shown in Figure 10. For fixed values of Reynolds number, the PDF is concentrated in the vicinity of unity in the upstream laminar region, and starts to broaden at the transition point, since the mixing is now provided mostly by the large scale fluctuation. In the subsequent transitional region, the PDF distribution changes very rapidly with Z to give a very broadened Gausian-like distribution ($Z = 8d$). In the further downstream region, the mixing through small scale turbulence comes into play and the distribution

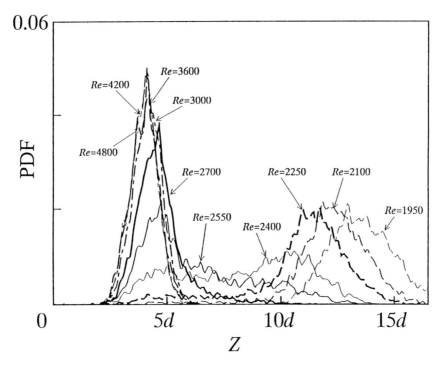

Figure 7 PDF distribution of transition length for fixed values of Reynolds number.

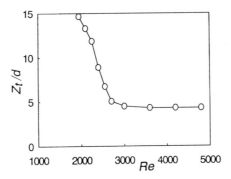

Figure 8 Dependence of transition length on Reynolds number.

becomes more concentrated with the peak location being shifted towards leaner side. When *Re* is increased, the PDF distribution in the downstream region tends to become more concentrated. This may suggest that the role of small scale turbulence in the mixing becomes more and more significant.

The above described general behavior of mixing processes in the jet can be studied further in terms of a quantity called scalar dissipation rate χ (Dahm, 1991). The scaler dissipation rate is defined by the following equation of the two-dimensional *z–r* plane and represents some measure of the degree of the molecular diffusion proceeding at each position in this plane.

$$\chi = D\left\{ \left(\frac{\partial C}{\partial z}\right)^2 + \left(\frac{\partial C}{\partial r}\right)^2 \right\}$$

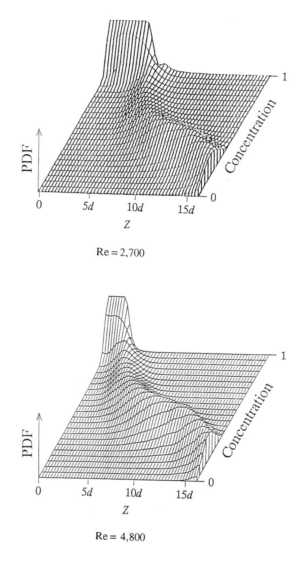

Figure 9   Concentration PDF distribution for Re = 2700 and 4800.

where C and D represent normalized jet fluid concentration and diffusion coefficient, respectively. χ is made nondimensional by constands D and d. An additional experiment was made to study the instantaneous distribution of this quantity. Figure 11 shows the enlarged instantaneous concentration field in the vicinity of the injector exit and the corresponding scalar dissipation rate field for Re = 5400. The color scale for χ is logarithmic so as to show the field in a wide range of the scalar dissipation rate. In the upstream laminar region, the string-like very concentrated region of high dissipation rate appears at the jet boundary suggesting that the molecular diffusion proceeds very rapidly there. It should be noticed that the size of the string is on the order of pixel size, and represents the smallest scale possible to visualize by the present method. The actual dissipation rate will be concentrated in narrower regions. In the subsequent downstream transitional region the string starts to fluctuate. The molecular diffusion proceeds mostly at the boundary of the fluctuation, suggesting that this large scale fluctuation is responsible for the mixing in this region. Further downstream, the string-like concentrated region of high dissipation rate

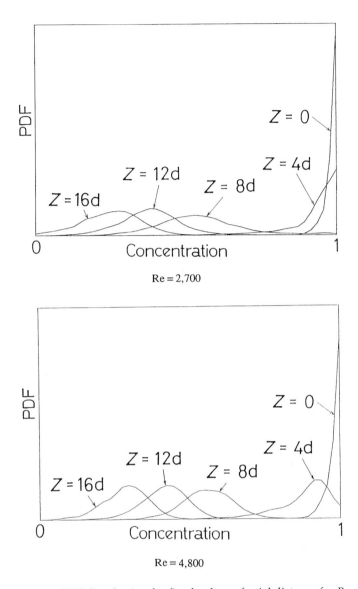

Figure 10   Concentration PDF distribution for fixed values of axial distance for $Re = 2700$ and 4800.

gradually disappears. Now, the small scale turbulence comes into play to get rid of the concentrated region where the molecular diffusion plays a role.

Concluding Remarks

The present experimental study has brought about the following concluding remarks:

1. The transition from laminar to turbulent flow in the jet is caused by a helical mode Kelvin-Helmholtz instability.
2. The mixing process in the transitional region, downstream of transition point, is mostly governed by the large scale fluctuation induced by the transition.
3. In the mixing process in the further downstream turbulent region, the role of small scale turbulence becomes more important.

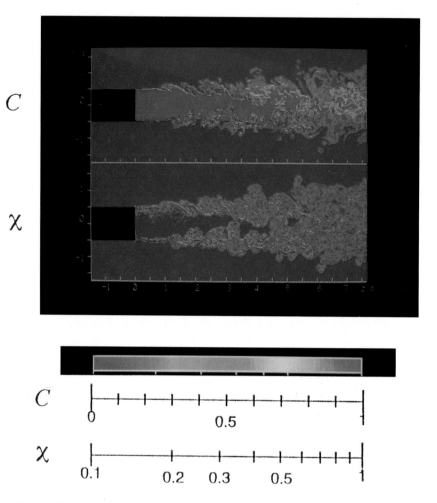

Figure 11 Enlarged instantaneous concentration field with corresponding scalar dissipation rate field.

Acknowledgment

The authors would like to express their sincere thanks to M. Kobayashi, Y. Machida, and N. Gotoh for their help in conducting the experiment.

References

Batchelor, G. K., 1958: Small-scale variation of convected quantities like temperature in turbulent fluid, Part 1 and Part 2, *J. Fluid Mech.*, Vol. 5, pp. 113–119.

Dahm, W. J. A., Southerland, K. B., and Buch, K. A., 1991: Direct, high resolution, four-dimensional measurements of the fine scale structure of Sc > 1 molecular mixing in turbulent flows, *Phys. Fluids*, A, Vol. 3, Part 5, pp. 1115–1127.

Kushida, G., Yamashita, H., and Takeno, T., 1993: A numerical study on transition and large-scale mixing processes in a coaxial jet, *Proc. 9th Symp. Turbulent Shear Flows*, Vol. 1, pp. 9-1-1 to 9-1-6.

chapter four

Two-Dimensional and Three-Dimensional Imaging of Passive Scalar Fields in a Turbulent Jet

G. J. Merkel[2], T. Dracos[1], P. Rys[2]

[1]Institute of Hydromechanics and Water Resources Management, [2]Dept. of Industrial Chemistry and Chemical Engineering, Swiss Federal Institute of Technology, Zurich

Abstract — Laser induced fluorescence (LIF) is used to visualize a passive scalar in a fully developed turbulent jet of moderate Reynolds number. Good resolution 2-D images of a longitudinal section and cross-sections of the jet at the macroscale level are presented. 3-D pictures of the structure of the scalar field at the level of Taylor's microscale are obtained by a tomographic method of high spatial and time resolution. This method opens new perspectives for the study of the development of scalar fields in laminar and turbulent flows in space and time and possibly for reactive flows as well.

Introduction

When chemical reactions are fast in relation to mixing processes, a significant conversion takes place before the reactants are completely mixed. In this case mixing processes can significantly affect the kinetics and product distribution in competitive, consecutive reaction systems (Bourne et al., 1981, Rys, 1992). Also the yield and selectivity of chemical reactions in fluids depend strongly on the rate at which the reacting partners are mixed (Wehrli et al., 1991).

One of the most efficient mixing processes in fluids is turbulence. In a homogeneous fluid the reactions take place at the interfaces between the mixing reactants. These interfaces are stretched and convoluted, and thus greatly increased, by turbulent eddies of all sizes. Therefore, it is of interest to visualize these interfaces and to determine concentration gradients on them. If in a first step chemical reactions are neglected, a passive fluorescent tracer can be used to mark one part of the mixing fluid for this purpose provided the Schmidt number of the tracer is large enough. Illumination by a sheet of laser light with the appropriate wavelength allows one to visualize the marked fluid by inducing the fluorescence of the dye.

This technique has been applied for some time by different investigators (for example, Breidenthal, 1981, Koochesfahani and Dimotakis, 1986, Papanicolau and List, 1988, Papantoniou and List, 1989, Dahm and Dimotakis, 1990), however it is mainly used for 2-dimensional observations. In the present work, the LIF technique is further developed along the line of Prasad and Sreenivasan (1990), Dahm et al. (1991) and Merkel et al. (1993) to a "Flow Tomography" by laser induced fluorescence (FTLIF) and used to observe 3-D volume of linear dimension of the order of the Taylor microscale with a resolution smaller than the Kolmogorov length scale. The investigation was conducted in a round, nonbuoyant jet as a mixing turbulent flow. The overall structure of the jet is visualized.

Experimental Facility

To enable observations of the jet flow over long periods of time the apparatus in which the jet was discharged had to fulfill certain conditions. In general, the recipient tank has limited dimensions. On the one side, this limits the duration of the experiments to the time the jet reaches the downstream boundary, where it eventually reflects. On the other side, the entrainment of the jet induces, in a tank of limited size, a more or less strong recirculation of the ambient fluid, which affects the jet behavior. A setup as shown schematically in Figure 1 and in the photo in Figure 2 was adopted to overcome these difficulties.

The main feature of this setup is a tank with the dimensions L = 3.0 m, W = 1.5 m and D = 1.0 m, which serves as the recipient of the ambient fluid. On one side the tank is equipped with a glass window of 1.0 m × 2.0 m, allowing illumination and observations of the flow. The jet is discharged at the center of a narrow side of the tank. Along the four corners of the long sides of the tank the estimated entrainment is supplied by four perforated pipes covered with filter material.

At the downstream end of the tank the total amount of fluid introduced in the tank is withdrawn through an array of perforated pipes and a constant head vessel. The fluid can either be recirculated by flowing back into the pump-swamps of the jet and entrainment supply systems, or directly discharged into the drainage system of the laboratory. The

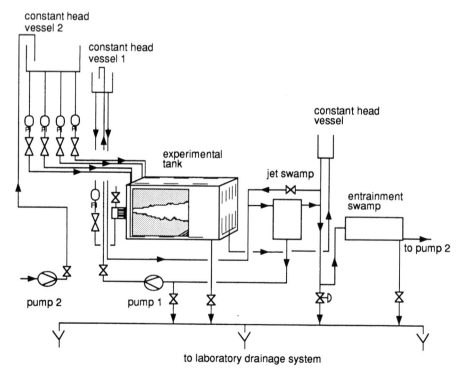

Figure 1   Schematic representation of the experimental facility.

Figure 2   Photo of the experimental facility.

latter is the case when the water in the tank becomes contaminated by the fluorescent tracer used to visualize the flow. The jet and the entrainment supply are controlled by constant head vessels and separately measured. The dye is injected by a volumetric pump in the duct supplying the jet fluid. Special care is taken to achieve a complete mixing of the dye with the fluid.

The jet issues from an orifice of 5 mm diameter mounted on a cylindrical plenum with an inner diameter of 150 mm. The centerline velocity u_m of a round jet is given by (Schlichting, 1965, Rajaratnam, 1976)

$$u_m = C_1 I_o^{1/2} x^{\pm 1} = C_1 u_o A_o^{1/2} x^{\pm 1} \tag{1}$$

I_o is the kinematic momentum flux, u_o is the velocity of the jet at the orifice, A_o is the cross-sectional area of the orifice, and x is the axial distance from the virtual origin. For $x / A_o^{1/2} \gg 1$, x can be taken equal to the distance from the orifice.

The half-width $b_{1/2}$ of the jet, i.e., the distance from the axis of the jet to the point at which $u/u_m = 0.5$ is found to be $b_{1/2} \sim 0.085x$ (Rodi, 1982), and the distance from the axis to the edge of the jet is $\delta \sim 0.22x$ (Dahm & Dimotakis, 1990). With $C_1 \sim 7.2$ we finally get

$$\frac{u_m}{u_o} = 6.38 \frac{d_o}{x} \tag{2}$$

where d_o is the diameter of the orifice.

The Reynolds number of the fully developed jet $(x/d_o \gg 40)$ is then

$$Re = \frac{u_m \delta}{\nu} = 1.4 \frac{u_o d_o}{\nu} = 1.4 \ Re_o \tag{3}$$

The entrainment Q_E is estimated to be (Schlichting, 1965)

$$Q_E = 0.404\ u_o A_o^{1/2} x = 0.358\ u_o d_o x \tag{4}$$

Next, an estimate of the viscous length scale λ_v and the diffusion length scale λ_D with Sc as the Schmidt number is given.

$$\lambda_v = \delta\ Re^{\pm 3/4} = 0.1706\ Re_o^{\pm 3/4} x \tag{5}$$

$$\lambda_D = \lambda_v Sc^{\pm 1/2} \tag{6}$$

The parameters of the jet in which the observations were made are given in Table 1.

Table 1   Parameters of the Experiment

Orifice diameter	$d_o = 5$ mm
Flow velocity at orifice	$u_o = 0.85$ m/s
Reynolds number at orifice	$Re_o = 4275$
Reynolds number of the jet	$Re = u_m \delta_v^{-1} = 6000$
Schmidt number	$Sc = \nu/D = 2075$
Location of observation plane	$x/d_o = 280$
	$r/b_{1/2} = 1.18$
$u_m\ (x/d_o = 280)$	19.4 mm/s
Viscous length at $x/d_o = 280$	$\lambda_v = \delta Re^{-3/4} = 0.45$ mm
Turn-over time	$t_\tau = \lambda_v^2/\nu = 0.2$ s
Exposure time per frame	$t_f = 0.0021$ s
Record time for a volume (50 frames)	$t_e = 0.105$ s

Principle of Tomography by LIF

Consider a turbulent part of a fluid being in contact with a nonturbulent part of the same fluid. The large eddies present in the turbulent part will engulf the nonturbulent fluid and expose it to the action of even smaller eddies, which will mix it into the turbulent part until it also becomes turbulent. To visualize this process, a fluorescent tracer with high Schmidt number can be introduced into one of the two parts of the fluid. Illumination by a thin sheet of, preferably, monochromatic light with a wavelength which is absorbed by the fluorescent tracer allows one to observe the turbulent mixing process in the illuminated plane. If this plane is rapidly moved parallel to its initial position, simultaneously recording pictures of the observed planes, a tomographic, quasi-instantaneous 3-D image of the mixing process can be constructed. The fluorescence of the tracer should be excitable at the wavelength of the light source used. The following conditions must be observed in order to fulfill the requirements of quasi-instantaneous tomographic observations.

- The spacial resolution normal to the observation plane should be of the same order as the resolution in the plane. The latter depends on the number of pixels in the rows and columns of the sensor used for imaging and on the imaging scale. The thickness of the lightsheet should be the same as the length of a line element in the object space imaged on a pixel.
- The time required for the observation of a 3-D element of the size of the smallest turbulent structure present in the flow should be significantly smaller than the turnover time t_v of such a structure.

$$\lambda_v = \lambda_v/u_\tau \tag{7}$$

λ_v is of the size of the smallest turbulence element, and u_τ is the friction velocity. It follows that:

1. The exposure time t_f for imaging a single observation plane must be $t_f < t_v$.
2. The time t_p required to image the number of consecutive observation planes necessary to cover the length λ_v in a direction normal to these planes must meet the condition $t_f < t_p < t_v$.

To fulfill these conditions the resolution and the imaging frequency of the camera must be as high as possible.

Illumination and Imaging for Flow Tomography

In the present work we seek a resolution to scales smaller than the Kolmogorov length scale. An EG&G Reticon camera, model MC4256 is used. The sensor of this camera consists of a 256×256 matrix of photodiodes with center to center distance of 39 µm and a sensor size of 10 mm × 10 mm. In the object-plane the observation area is 15 mm × 15 mm, or 33 $\lambda_v \times 33\ \lambda_v$. The linear dimension of this area is of the order of Taylor's microscale. In object space, the spatial resolution achieved with this camera is 60 µm or 0.13 λ_v.

The camera can operate at a maximum image rate of 500 Hz in both a fixed rate mode or with external triggering. The high imaging frequency is achieved by high-speed read-out through 8 outputs. Four parallel A/D converters digitize the data and store them in a 64 MB memory card. This allows one to store a sequence of 1024 consecutive images. The camera system offers high flexibility for scientific image data acquisition but the images show a pattern of vertical stripes. Therefore an algorithm was developed to destripe the images (Maas, 1993).

An Argon ion laser (Coherent Innova 420) is used as light source. Its total light power is 26 W with 10 W concentrated in the 514.5 nm line and 8 W concentrated in the 488 nm line. The laser beam has a diameter of 1.9 mm. To achieve the same resolution in all three directions this diameter must be reduced to an average of 60 µm in the range of the imaged plane. For this purpose the following optical arrangement is used: first a beam expander expands the beam by a factor of 16 to a diameter of approximately 30 mm; the beam is then focused on the observation location in the jet by an adjustable focusing lens system. This optical arrangement is in line with the laser beam, which is horizontal and parallel to the glass window of the tank. The focused beam is deflected upwards in vertical direction by a fixed mirror. At the desired elevation a second mirror deflects the beam in a direction perpendicular to the axis of the jet. The laser beam is then expanded by a cylindrical lens to form a sheet of light of a mean thickness of 60 µm. The second mirror is mounted on a piezo scanner unit (Figure 3). This unit has a maximum angle of deflection $\alpha = 2.25$ mrad, which can be subdivided in 4096 steps with a maximum stepping frequency of 5 kHz. In the present application a range of 2.23 mrad was divided into 50 steps of 44.6 µrad. The mirror was moved with a frequency of 486 steps/s. After 50 steps performed in 0.105 s, the mirror is moved back to its initial position, and the procedure is repeated. The storage capacity of the camera system allows one to perform 20 repetitions in a total time of 2.10s. Each step of the mirror produces a displacement of the light sheet at the observation location by 60 µm. This way an observation volume of 15 mm × 15 mm × 3 mm is created each 0.105 s, with a resolution of 60 µm in all three directions (Figure 4).

The local mean of the center line velocity of the jet at a distance of 280 d_o from the orifice where the observations were made is approximately 19.5 mm/s. In lateral direction, the flow was observed at a distance $r/b_{1/2} = 1.18$ from the axis of the jet. At this location the mean jet-velocity is reduced to approximately 7.5 mm/s. To correct for the advective

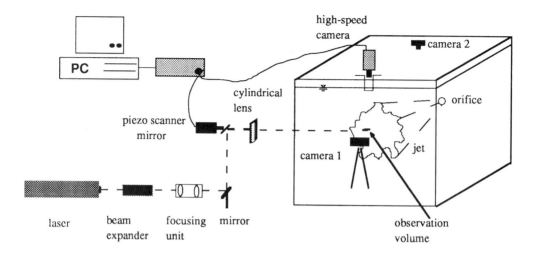

Figure 3 Scheme of the optical arrangement.

displacement of the fluid during the imaging of the volumes the camera and the optical arrangement are mounted on a carriage, which can be moved by a programmable stepper motor with the desired velocity, in our case with a velocity of 7.5 mm/s. The camera mount has three linear displacement units for precise positioning in x, y, and z directions. Here x is parallel to the axis of the jet, positive in flow direction, y is horizontal, and z is vertical forming a right-handed coordinate system. To avoid optical distortions due to small perturbations of the free water surface in the tank a cylinder with a glass window submerged 2 to 3 cm in the fluid was mounted in front of the camera.

The stepwise movement of the mirror and the camera are synchronized by using the derivative of the stepwise increasing voltage controlling the movement of the mirror to externally trigger the camera.

Overall Imaging by LIF

To gain information about the overall structure of the jet, a macroscale imaging of the jet was performed in addition to the tomographic observation of the jet structure at the microscale level. For this purpose the optical arrangement was adapted as follows:

- The cylindrical lens was put in line immediately after the focusing lens system.
- The two reflecting mirrors were replaced by two large mirrors. The first had the dimension 150 mm × 700 mm and the second 150 mm × 980 mm. These mirrors were

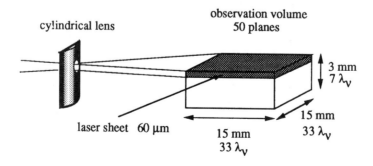

Figure 4 Tomographic generation of an observation volume.

arranged in such way that either a horizontal sheet of light passing through the axis of the jet or a vertical sheet of light normal to the axis of the jet could be produced. The thickness of these sheets was 0.5 mm and their width at the axis of the jet was approximately 600 mm.

- Images were taken by a Nikon AF motor driven camera at a rate of two images per second. The exposure time varied between 1/250 s and 1/500 s.
- The camera was mounted on the carriage with the observation cylinder mounted in front of the lens when horizontal cuts through the jet were imaged. For taking cross-section views, the camera had to be placed with its optical axis aligned with the axis of the jet. For this purpose, a water-tight aluminum container with a glass window was constructed and placed at the downstream end of the tank. For focusing the camera, the readout of the eyepiece of the camera was imaged by a CCD camera and displayed on a monitor.

Overall Structure of the Jet

The experiments were performed in the following way: the tank and the water storage system were filled with tap water. A long enough time was allowed for the system to adapt to the ambient temperature. The jet and the entrainment supply system were then started and the discharges adjusted. The jet was left to run for a few hours to further homogenize and stabilize the fluid temperature in the whole system. The dye Na-Fluoresceine (Uranin) was then injected in the water supply of the jet, and shortly thereafter the recirculation stopped. Na-Fluoresceine has a maximum absorbance at 493 nm and emits at 513 nm. After an initial period of time the jet was colored over its entire length. At this point imaging was started.

Figure 5 shows a horizontal axial cut through the jet. The exposure time is 1/250 s. The spreading of the jet is symmetric and the total spreading angle is approximately 26° as expected. The length imaged extends from $x/d_o = 0$ to $x/d_o = 120$. The internal structure

Figure 5   Longitudinal cut through the axis of the jet. Thickness of the light sheet 0.5 mm.

of the jet is very similar to the one observed by other authors at Reynolds numbers of the same order of magnitude (e.g., Dahm and Dimotakis, 1990). It is typical for a fully turbulent jet.

The cross-section of the jet was imaged at a distance $x / d_o = 240$ from the orifice. Figure 6a and b shows the structure at two different times 5 s apart. The exposure time was $1/500$s. The imaging scale is, in both cases, the same. A black asterisk marks the axis of the jet in Figure 6a and b. Large patches of dark ambient fluid are entrained up to the axis of the jet. Note, however, that unmixed ambient fluid is also found in thin streaks inside the small-scale structures all over the cross-sectional area of the jet.

Square dots in Figure 6b mark the relative position at which tomographic observations are made. The maximum width of the cross-section in Figure 6b is approximately 52.5 mm in object space. In the scale of this picture the extend of the part imaged in the tomographic observations is indicated at the lower right corner of Figure 6b by a white bar. It is of the order of the small-scale structures discernible in the photographs of the cross-section of the jet.

Tomographic Observations

The imaging system is described in Chapter 3. Note that the observation is made in a Lagrangian frame, i.e., the optical system and the camera are moved with the local mean velocity at the observation site. The camera has a 10-bit grey scale resolution, however, the fast framing processor board connected to the camera can only resolve eight bits providing 256 shades of grey.

In the false color mapping the 256 shades of grey are linearly represented by the color scale from red to blue in decreasing grey level. Hence, red indicates jet fluid with a high concentration of dye and blue unmixed ambient fluid. Figure 7 shows five consecutive tomographic observations of a fluid volume with dimensions of 15 mm × 15 mm × 3 mm. Each of these volumes is composed of 50 individual images of a 15 mm × 15 mm plane. The thickness of the light-sheet for each image is 60 μm on the average and each image is displaced in a direction normal to the imaging plane from top to bottom by 60 μm. Three sides of the volume are visible in the false color pictures of Figure 7 giving a 3-D portrayal of a turbulence structure and its evolution with time. The size of this structure is of the order of Taylor's microscale. Inside the structure the rollup of the interface between jet fluid and ambient fluid is visible. Though a sub-Kolmogorov scale resolution is achieved, no smaller scale eddies are discernible. However, the streaks of different colors often have a thickness of the order of the viscous length.

These images are now being analyzed in different ways. Attempts have been made to determine 3-D velocity fields by using sequences of tomographic imaging of marked fluid volumes (Mass, 1993). Here we present the 3-D gradient analysis of grey levels as well as the 3-D scalar dissipation in a volume. Gradients were computed by the "nearest neighbor" algorithm for the distribution of scalar $\zeta(x, y, z, t)$ in the volume shown in Figure 8a. Figure 8b shows the computed gradients in the midplane of the volume. Scalar dissipation $\nabla\zeta \cdot \nabla\zeta(x, y, z, t)$ was computed and is depicted in Figure 8c. Dissipation of the scalar is taking place in 3-D thin sheets of thickness comparable to the viscous length scale. The analysis of the synoptic and the tomographic images is in progress and will be published in due time.

Summary

Tomography by LIF can be used to study the structure of turbulent flows at moderate Reynolds numbers. This method allows one to resolve scales down to the Kolmogorov length scale in flows of moderate Reynolds numbers. The time resolution is so high that

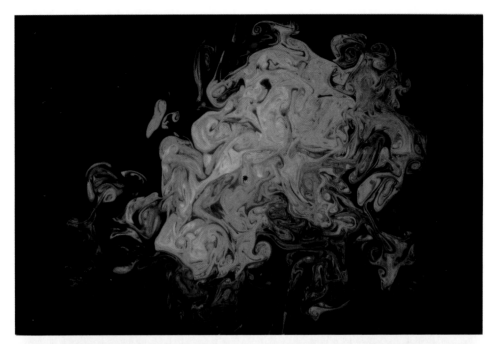

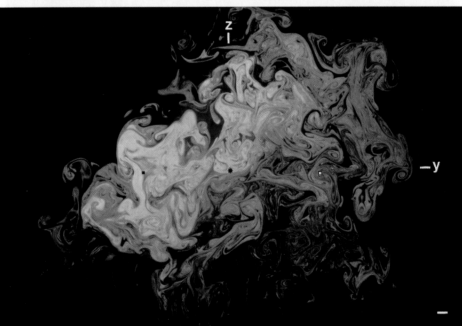

Figure 6 Two cross-sectional images of the jet taken in a time interval of 5 s. *: Location of the axis of the jet; white square: relative location of the tomographic observation; white bar: size of the field imaged tomographically.

the tomographic image of a moving fluid element of linear dimension comparable to Taylor's microscale can be considered as quasi-instantaneous. The time interval between consecutive imaging of tomographically observed volumes is also smaller than the turn-over time of the smallest structures. A sequence of tomograms thus gives one the possible ability to study the evolution of the small-scales in mixing turbulent flows.

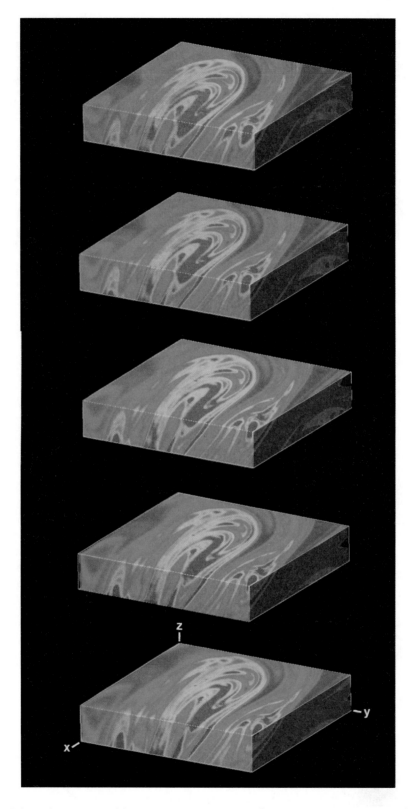

Figure 7   False color images of five consecutive tomographically observed volumes showing the distribution of dye concentration. The x-axis shows in flow direction. Red indicates high dye concentration: blue ambient fluid with zero dye concentration.

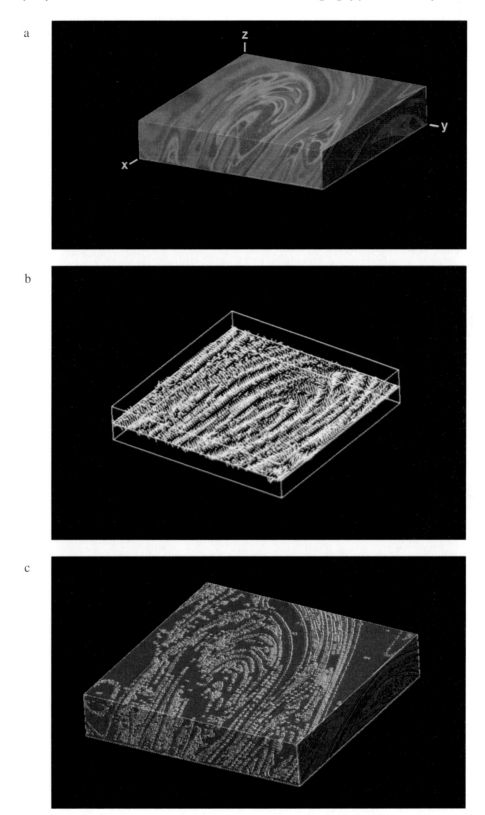

Figure 8a, b, and c   (a) False color distribution of the concentration of the scalar $\zeta(x, y, z, t)$; (b) Scalar gradient $\nabla\zeta(x, y, z, t)$ at the midplane of the observation volume; (c) Logarithm of the scalar dissipation log $\nabla\zeta \cdot \nabla\zeta(x, y, z, t)$. Red: maximum dissipation; blue: minimum dissipation.

Acknowledgments

We would like to thank Mr. K. A. Buch, from the Department of Aerospace Engineering, University of Michigan, for developing the computer programs. The work presented here is being supported by research grant No. 41-2516.5 of the Swiss Federal Institute of Technology, Zurich.

References

Bourne, J. R., Kozicki, F., Moergeli, U., and Rys, P., 1981: Mixing and fast chemical reaction-III. *Chem. Eng. Sci.*, 36, 1655–1663.

Breidentahl, R., 1981: Structure in turbulent mixing layers and wakes using a chemical reaction. *J. Fluid Mech.*, 109, 1–24.

Dahm, W. J. A. and Dimotakis, P. E., 1990: Mixing at large Schmidt number in the self-similar far field of turbulent jets. *J. Fluid Mech.*, 217, 299–330.

Dahm, W. J. A., Southerland, K., and Buch, K. A., 1991: Direct high resolution four dimensional measurements of the fine scale structure of $Sc \gg 1$ molecular mixing in turbulent flows. *Phys. Fluids*, A3, 1115–1127.

Koochesfahani, M. M. and Dimotakis, P. E., 1986: Mixing and chemical reactions in a turbulent liquid mixing layer. *J. Fluid Mech.*, 170, 83–112.

Maas, H.-G., 1993: Destriping of digital images. *ISPRS Workshop on Digital Sensors and Systems*, Trento, Italy.

Maas, H.-G., 1993: Bestimmung dreidimensionaler Geschwindigkeitsfelder aus Strömungstomographiesequenzen. 15. *DAGM Symposium Mustererkennung*, Lübeck, Germany.

Merkel, G. J., Dracos, T., Rys, P., and Rys, F. S., 1993: Flow tomography by laser induced fluorescence. *XXV IAHR Congress*, Tokyo, Japan.

Papanicolau, P. N. and List, J. E., 1988: Investigations of round vertical turbulent buoyant jets. *J. Fluid Mech.*, 195, 341–391.

Papantoniou, D. and List, J. E., 1989: Large-scale structure in the far field of buoyant jets. *J. Fluid Mech.*, 209, 151–209.

Prasad, R. R. and Sreenivasan, K. R., 1990: Quantiative three-dimensional imaging and the structure of passive scalar fields in fully turbulent flows. *J. Fluid Mech.*, 216, 1–34.

Rajaratnam, N., 1976: *Turbulent Jets*. Elsevier, Amsterdam.

Rodi, W., 1982: *Turbulent Buoyant Jets and Plumes*. Pergamon Press, Oxford.

Rys, P., 1992: The mixing-sensitive product distribution of chemical reactions. *Chimia*, 46, 469–476.

Schlichting, H., 1965: *Grenzschicht Theorie*. G. Braun, Karlsruhe.

Wehrli, M. B., Rys, P., and Rys, F. S., 1991: Numerical analysis of mixing and chemical reactions in a two-dimensional laminar flow. *Proc. Int. Workshop Chem. React. Turbulent Liquids*, Lausanne, Switzerland.

chapter five

The Use of Liquid Crystals and True-Color Image Processing in Heat and Fluid Flow Experiments

J. Stasiek[1] and M. W. Collins[2]

[1]Technical University of Gdansk, Poland, [2]Thermo-Fluids Engineering Research Centre, City University, London, U.K.

Abstract — *This paper describes noninvasive methods which can determine quantitatively two-dimensional (2-D) temperature distributions on a surface and in a fluid from color records obtained using a thermosensitive liquid crystal material combined with image processing. Application-type experiments have been carried out both to visualize the complex temperature distribution over a cooled surface disturbed by different solid obstacles, and also to investigate temperature and flow patterns in a rectangular cavity for natural convection.*

Introduction

The theoretical equations for fluid flow and convective heat transfer are complex, nonlinear partial differential equations requiring the simultaneous satisfaction of continuity, momentum, and energy principles. Current standard computational fluid dynamics (CFD) codes can now deal with three-dimensional (3-D) transient problems associated with irregular geometries. Using the largest current computers, it is now possible to directly treat not only 3-D laminar transient flows, but also transitional and low turbulent flows. For the latter two, the procedure is termed *direct simulation*. For high turbulence, however, the movement of the larger, slower eddies may be resolved (large eddy simulation or LES) while the unresolved smaller eddies require a subgrid model.

The whole-field nature of such numerical treatments, and the broader question of general code validation, demand corresponding experimental methods. (Laser doppler anemometry and more recently particle image velocimetry (PIV) have been established and are extensively used for obtaining point velocities and planar velocity field measurements respectively in wide ranging laboratory and industrial situations.) Fluid temperature measurement techniques are usually invasive and difficult near a wall. Holographic interferometry, on the other hand, is another of the number of new noninvasive optical methods. It gives whole-field instantaneous data on a par with LES and can be applied both qualitatively and quantitatively for thermal structure visualization.

To understand such thermohydrodynamic phenomena, it is very informative to visualize the detailed distribution of the surface temperature and temperature and velocity fields in liquids.

The purpose of this paper is to report a new technique that allows visual determination of both qualitative and quantitative heat transfer and fluid flow in thermofluid investigations. Moffat (1991) and Jones et al. (1992) have recently reviewed available methods of heat transfer/temperature measurement and their applications. Thermochromic liquid crystals (TLC) are identified as a convenient measurement method offering satisfactory accuracy and resolution. Also, during the past several years, there has been an increasing use of computers, automatic recording, and fringe and data processing software in heat transfer laboratories; first, to facilitate conventional experiments and second, to expand the range of possible work. Today, due to a synergetic combination of thermochromic liquid crystal and fast, small computers, we are seeing what amounts to a revolution in heat transfer and fluid flow research. With the added advantage of automatic interpretation using image processing techniques, liquid crystal thermography is a very powerful, yet simple, measurement tool for the determination of quantitative heat transfer data. It is one of a range of new noninvasive, whole-field optical measurement methods.

In this paper we carefully review the above issues, and use illustrative examples from our own work in applying TLC to the study of forced and natural convective heat transfer, first on a cooled surface heated by air flow disturbed by a number of complex geometrical configurations, and also for temperature and flow visualization in rectangular cavity, respectively.

Liquid Crystals: A Background Survey

In 1988 an Austrian botanist, Fredich Reinitzer observed that certain organic compounds appeared to possess two melting points (mp), an initial mp at which the solid phase turned to a cloudy liquid and a second mp at which the cloudy liquid turned clear. Further research revealed that an intermediate phase or *mesophase* did, indeed, exist between the pure solid and pure liquid phases of some organic compounds. Reinitzer termed this phase *liquid crystal*, an appropriate designation when one considers that the material exhibits the fluidity of a liquid while maintaining the characteristic of the anisotropic, ordered structure of a crystalline solid. The mesophases have been classified as nematic, cholesteric, smectic, and blue; Jones et al. (1992) and Parsley (1988). However, from a structural viewpoint, it can be argued that there are only two basic types of liquid crystal; chiral-nematic and smectic, with cholesteric being regarded as a special kind of nematic. Their structures are shown schematically in Figure 1. A question of great relevance is the texture taken up by the cholesteric phase which can exhibit two forms; the focal-conic and the

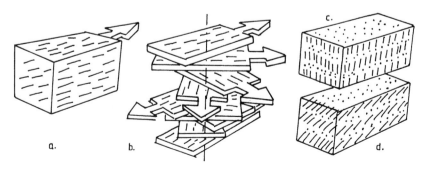

Figure 1   Mesophase structures: (a) nematic; (b) Cholesteric; (c) Smectic A; (d) Smetic C (Jones et al. 1992).

grandjean textures. In the focal-conic texture the cholesteric helices pack around elliptical and hyperbolic paths, whereas in the grandjean form the helical axes are in the same direction and large regions of aligned molecules are present.

This texture is birefringent but optically inactive. In the grandjean texture, which forms from the focal-conic by mechanical shear, the helices are more or less all aligned with their axis parallel to incident light. This is the texture that exhibits the unique optical effects of the mesophase-iridescence, optical activity, circular dichroism, and selective reflection of white light to show brilliant color. Over a known, reproducible range of temperature, the *event temperature range*, the cholesteric liquid crystal will progressively exhibit all colors of the visible spectrum as it is heated through the event temperature range. The phenomenon is reversible, repeatable, and, with proper care, color can be accurately calibrated with temperature. Hence thermal transients are measurable.

Both the color play interval (that is, the range of temperature over which a certain color is visible) and the event temperature range of a liquid crystal can be selected by adjusting its composition, and materials are available with event temperatures from –30° to 120°C and with color play bands from 0.5 to 20°C, Jones et al. (1992) and Moffatt (1991), although not all combinations exist of event temperature and color play band widths. Widths of 1°C or less will be called *narrow band* materials, while those whose band width exceeds 5°C will be called *wide band*. The types of material to be specified for a given task depend on the type of image interpretation technique to be used.

Pure liquid crystal materials are thick, viscous liquids, greasy and difficult to deal with under most heat transfer laboratory conditions. Two approaches have been used to make them more practical to use:

1. encapsulation of the liquid crystal in a gelatine-like material forming nearly spherical particles from 10-30 microns in diameter and then making a paint using those particles as the pigment, or
2. application of the unencapsulated material (unsealed liquids) to a clear plastic sheet and sealing it with a black backing coat to form a prepackaged assembly. This is because in their (pure) exposed state TLC quickly degrade.

Calibration of Liquid Crystals and Image Processing

Before the execution of a thermal or flow visualization experiment, we should recognize the characteristics of the overall combination of the TLC, the light source, the optical, and camera system (ordinary, CCD black-and-white, or RGB color camera), and make a rational plan for the total measurement system. Figures 2 and 3 show the apparatus of a calibration experiment. A known temperature distribution exists on a *calibration plate* to which the liquid-crystal layer is attached.

This experiment was also designed to investigate viewing angle effects. In order to maintain a linear temperature distribution with desired temperature gradients, one end of a brass plate was cooled by stabilized water and the other end electrically controlled to give a constant temperature. The distribution of the color component pattern on the liquid-crystal layer was measured using a photo-camera with attached optical polarizer and also an RGB color camera. In the present study several temperature gradients were tested. Figure 4 shows the relationship between Hue (one of the chromaticity characteristics explained later) and temperature ($\Delta t / \Delta X = 94$ K/m) on an RGB color coordinate — there is a tendency for larger Hue values to occur for larger temperature gradients. In this particular experiment uncertainty was estimated at about ±0.075°C because the specimen was calibrated in place with the lighting and RGB camera system arranged just as it was during the application.

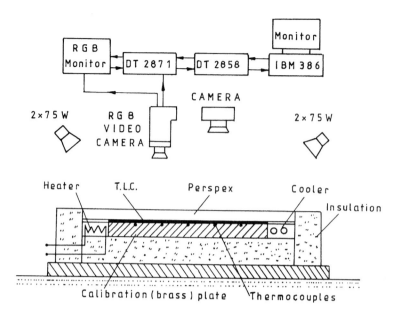

Figure 2   Experimental apparatus for the calibration experiment and true-color image processing interpreter system.

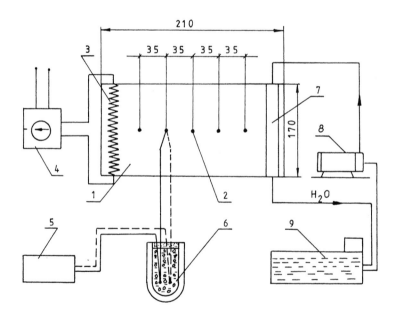

Figure 3   Schematic drawing of calibration (brass) plate circuit diagram: (1) brass plate; (2) thermocouples-type T; (3) heater-200W; (4) autotransfer; (5) digital voltmeter; (6) dewar flask with melting ice; (7) cooler; (8) water pump; (9) TE-8A constant temperature water water baths.

In the early days of image processing, only monochrome systems were available. Soon after, color image processing came into existence, but monochrome continued to dominate because it was easier and less costly to manufacture and less complicated to use. However, in the last four or so years, color image processing has gone from being used mainly in highly technical applications involving expensive image-processing systems to being available to virtually anyone who can use a desktop computer.

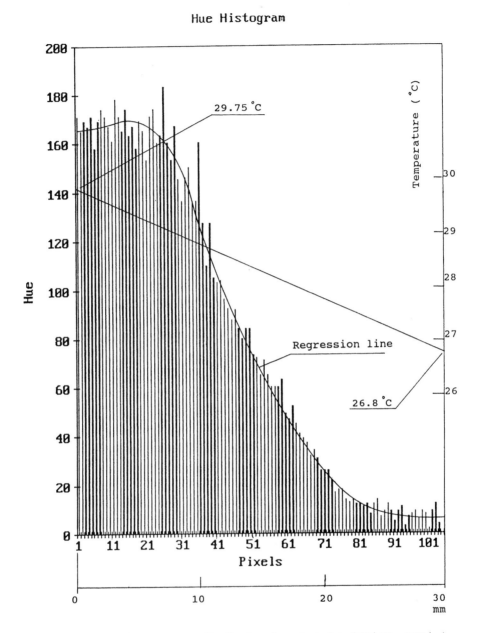

Figure 4   Hue and temperature distribution along test plate ($\Delta T/\Delta X = 94K/m$).

There are several commercial packages of hardware and software available which will digitize a video image from either black and white or color cameras and store individual frames pixel by pixel in arrays. Three separate arrays are used to store a color image, one each for the red, green, and blue signal components. Also, three styles of camera have been used; NTSC composite (the usual home video camera), CCD (charge-coupled device) and CID (charge injected device), Jones et al. (1992).

According to Akino et al. (1990), Moffat (1991), and Jones et al. (1992) there are four broad classes of image interpretation techniques available:

- human observers,
- multiple narrow-band spectral intensity image processing system,

- multiple linear regression method, and
- true-color image processing systems.

Human observers can interpret liquid-crystal images by direct visual inspection of color photographs or tape recorded video images, usually using narrow-band paints. Calibrations for such use are generally limited to identifying the temperature associated with a particular color, often the red or the yellow-green color, near the center of the color-play band. The uncertainty associated with direct visual inspection is about 1/3 the color-play band-width, given an observer with normal color vision and a little experience, Moffatt (1991), or about ±0.2 to 0.5°C, depending on the band-width.

The second and third methods are based on the interferential optical filters with narrow band-play characteristics of transmittance, through which equally colored regions can be extracted. Akino et al. (1989) developed a multiple narrow-band spectral intensity interpreter using a set of eighteen narrow band-pass filters, whose central wavelengths were from 400 to 750 nm, and the full width of the half maximum (FWHM) of the filters were less than 10 nm. The behavior of liquid crystals can be interpreted in terms of the variation of intensity with wavelength at a given temperature instead of the variation of dominant with Hue temperature. This approach gives rise to calibrations based on spectral intensity, as reported in Akino et al. (1989) — the resolution of this method is better than 0.1°C.

In the fourth method, color video information is displayed as a unique brightness combination of red, green, and blue (RGB) light. A displayed RGB image is immediately perceived by a user as exhibiting three distinct attributes:

1. **Hue** — the spectal colors present.
2. **Saturation** — how deep or faded the colors appear to be.
3. **Intensity** — the edge information or what would be seen if a black-and-white version of the new image were displayed. Hence, only intensity is available from a monochrome system.

Many image processing operations developed for processing grey-level (intensity) images can be readily applied to HSI color images. These include image transformations, enhancement, analysis, compression, transformations, and restorations. Building on gray-scale image processing hardware and software technology, Data Translation has produced a 512×512 pixels $\times 8$ bit (256 grey levels) color frame-grabber board for PC/ATs.

A schematic view of such an image processing system (developed by the authors) is shown in Figure 2. The two-dimensional temperature distribution is determined using an RGB video-camera, IBM 386 Personal Computer AT, HSI Color Frame Grabber DT 2871, and Auxiliary Frame Processor DT 2858.

Experimental Technique

Two main methods of measurement are performed involving steady state and transient techniques. A brief history of these methods is given in Baughn et al. (1989). Recent reviews of heat transfer measurements have also been produced by Moffatt (1991), Jones et al. (1992), and Ashforth-Frost et al. (1992).

Steady State Analysis — Constant Flux Method

The steady state techniques employ a heated model and the TLC is used to monitor the surface temperature. Usually, a surface electric heater is employed such that the local heat flux, $\dot{q}$, is known and this, together with the local surface temperature, T_w (found from the TLC), gives the local heat transfer coefficient, h:

$$q = I^2 r \text{ and } h = \frac{\dot{q}}{T_a \pm T_w}$$ (1)

T_a is a convenient driving gas temperature, I is the current, and r is the electric resistance per square of the heater.

Steady State Analysis — Uniform Temperature Method

The TLC-coated test specimen forms one side of a constant temperature water bath and is exposed to a cool/hot air flow. The resulting thermograph is recorded on film or video and further measurement positions are obtained by adjusting the water bath temperature. This method is more time consuming due to the large volume of water that needs to be heated. In this case, the heat transfer coefficient is determined by equating convection to the conduction at the surface:

$$h\left(T_a \pm T_w\right) = \frac{k}{x}\left(T_w \pm T_b\right)$$ (2)

where T_b is a water temperature, x the wall thickness, and k the thermal conductivity.

Transient Method

This technique requires measurement of the elapsed time to increase the surface temperature of the TLC-coated test specimen from a known initial temperature to a predetermined value. The rate of heating is recorded by monitoring the color change patterns of the TLC with respect to time. If the specimen is made from a material of low thermal diffusity and chosen to be sufficiently thick, then the heat transfer process can be considered to be one-dimensional (1-D) in a semi-infinite block. Numerical and analytical techniques can be used to solve the 1-D transient conduction equation. The relationship between wall surface temperature, T, and heat transfer coefficient, h, for the semi-infinite case is:

$$\frac{T \pm T_i}{T_a \pm T_i} = 1 \pm e^{\beta^2} erfc(\beta); \ \beta = h\left(\frac{t}{\rho ck}\right)^{0.5}$$ (3)

where, ρ, c, and k are the model density, specific heat, and thermal conductivity. T_i and T_a are the initial wall and gas temperatures and t is time from initiation of the flow, Baughn et al. (1989) and Jones et al. (1992).

Shear Stress Measurements

TLCs can also indicate shear stress and their use in neat (unencapsulated) form in this mode gives transition data. It was first investigated at NASA, Moffatt (1991), and a quantitative measurement method for skin friction was invented later by Bonnett et al. (1989). They observed that if a material in the focal-conic texture on a given surface is suddenly subjected to a steady air flow, then the time for the material to align into grandjean texture is a unique function of the shear stress. This time can be easily measured as it corresponds to the time for the surface to change from a colorless state. Thus, a robust technique independent of viewing angle or illumination became available. In this last case, transition was readily detected, Bonnett et al. (1989).

Temperature and Flow Visualization in Fluids

Liquid crystals have been used to make visible the temperature and velocity fields in fluids as well as in air. Temperatures have been visualized using the materials in the unsealed (neat) and microencapsulated forms as both tracer particles in flow field studies and as surface coatings, Hiller et al. (1986), Moffatt (1991), and Parsley (1988).

Application Experiments and Procedure

In order to demonstrate the feasibility of TLC techniques in practical heat transfer contexts, the authors have performed several experiments. The first set was carried out to investigate temperature and heat transfer coefficient distributions on a cooled surface heated by an air flow and distributed by a number of complex geometrical configurations, namely:

1. Crossed-corrugated geometrical elements as used in rotary heat exchangers (regenerators) for fossil-fuelled power stations.
2. Cylinders (both single and double) and a square section column. } relevant to local heat exchanger performance
3. Square roughness elements. or enhancement

   In the second set of experiments, flow structures and temperature distribution in a rectangular cavity were visualized using photographic records of the motion of tracer particles illuminated by a sheet of white light. Unsealed TLCs were used to make visible the temperature and velocity fields in a glycerol-filled cavity, which could be angled from *horizontal* through *inclined* to *vertical*.

Design and Construction of the Wind Tunnel

The experimental study was carried out using an open low-speed wind tunnel, Figure 5, consisting of an entrance section with a fan and heaters, large settling chamber with diffusing screen and honeycomb, and then mapping and working sections. Air is drawn through the tunnel using a fan able to give Reynolds numbers of between 500 and 50,000 in the mapping and working sections. The working air temperatures in the rig range between 25 to 65°C and are produced by the heater positioned just downstream of the inlet. The major construction material of the wind tunnel is perspex. Local and mean velocity are measured using conventional Pitot tubes and DISA hot-wire velocity probes.

   The wind tunnel is instrumented with copper-constantan (Type T) thermo couples and a resistance thermometer so that the surface, water bath, and air temperatures can be measured and controlled by a variac control system. In this experimental study the encapsulated thermochromic liquid-crystal layer is applied directly onto the cooled surface disturbed by the various shapes, Figures 6 and 7. The alternative effects of constant wall temperature and constant heat flux boundary conditions are obtained using a water bath, while the temperature can be controlled with a thermostat capable of establishing and maintaining temperature to within ±0.01°C accuracy. Photographs are taken using a standard camera, RGB video-camera, and true-color image processing system.

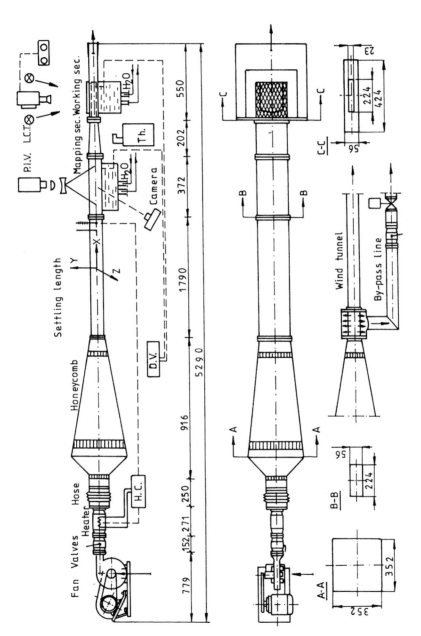

Figure 5 Open low-speed wind tunnel.

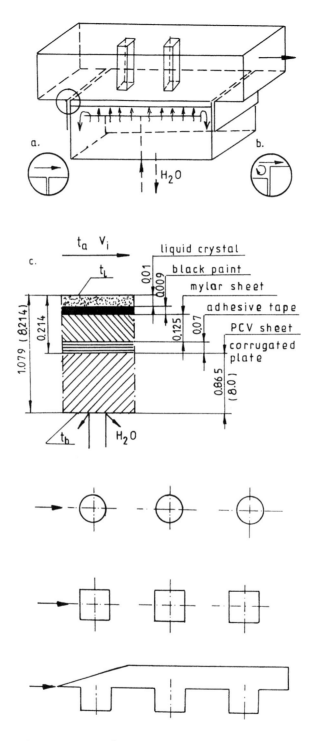

Figure 6   Mapping section geometry with liquid crystal package and scheme of three different end walls of cascade.

Experimental Set-Up for Flow Visualization in a Rectangular Cavity

Natural convection phenomena in rectangular cavities with differentially heated walls have received considerable attention due to their many applications in energy-sensitive

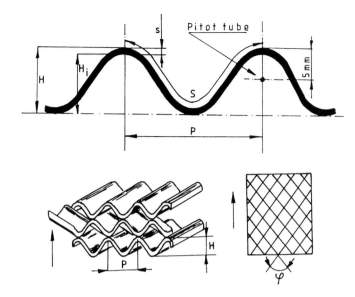

Figure 7   Three-dimensional corrugated geometry passage and cross-corrugated heat transfer element.

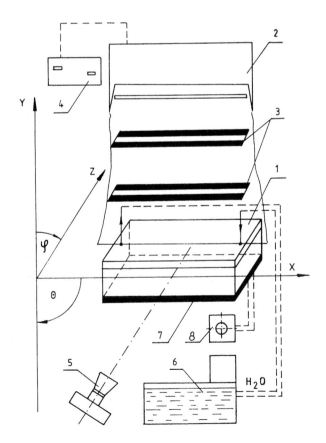

Figure 8   Schematic diagram of rectangular cavity facility: (1) cavity; (2) power flash lamp; (3) light slots; (4) programmable generator trigger; (5) camera; (6) water bath; (7) blanket 300W heater; (8) transform.

designs such as building systems with cavity walls and air gaps in unventilated spaces, double glazing, solar collectors, and furnaces. There are numerous other examples. These complex phenomena have given rise to a very large number of numerical predictive studies. However, quantitative fundamental experiments are still necessary to lead to better understanding of both the physical phenomena and their numerical simulations.

The thermal convection flow in this study was generated in a rectangular cavity of 180mm long, 60 mm wide, and 30 mm high, see Figure 8. The cavity had isothermal hot and cold walls and was made from 8 mm perspex (apart from the lower copper plate). The angle of the inclination of the cavity was varied from the horizontal to the vertical for two different planes. The photographs were made perpendicular to the line of the illumination and were taken with a 35 mm SLR camera with 90 mm lens.

The specimen is illuminated using an 800W xenon flash lamp collimated by a cylindrical lens and two slots to form a slit which may be adjusted over 2-3 mm in width. The flash lamp is triggered by a programmable impulse generator at a prescribed time sequence. Usually 4-12 flashes are used to take one color photograph. The time interval between flashes varied from one to several seconds. To obtain information about the direction of the flow, the last flash of a series has been released at half of the prescribed time interval, Hiller et al. (1986).

Thermochromic liquid crystals were used to make visible the temperature and velocity fields in the glycerol-filled cavity. The glycerol, liquid-crystal mixture was prepared by dissolving unencapsulated chiral-nematic materials (TM256, B.D.H. Limited) in ether and then spraying the mixture into the air above a free surface of glycerol. The ether evaporated in mid-air, leaving small drops of liquid-crystal material which fell into the glycerol forming an "almost monodispersed" suspension of particules approximately 50-80 microns in diameter. The concentration was kept below 0.03% by weight.

In the present experiments interest centered on the shape of the isotherms (not in their absolute temperature values) and on the possibility of applying a True-color image processing system for general observations of 3-D structures of heat and fluid flow in an enclosure.

Experimental Results

The heat transfer coefficient is a defined quantity, calculated from the surface heat flux and the difference between the surface temperature and some agreed reference temperature. This is usually the far field temperature, the mixed mean temperature, or the adiabatic surface temperature. In the following experiments liquid crystals were used to determine the distribution of surface temperature and heat flux — this allows for evaluation of the local heat transfer coefficient or Nusselt number. The temperature recorded from the liquid-crystal sheet is only at the red color (yellow-green for image processing), this being a well defined color for human viewing.

The local convective heat transfer coefficient h_ℓ in "ℓ" follows from the fact that the conductive heat flux q_k in either working section is equal to the convective heat flux from air to the surface in the stationary state:

$$h_\ell = \frac{k_t}{\delta_t} \frac{\left(T_\ell - T_b\right)}{\left(T_a - T_\ell\right)} \qquad \textbf{(4)}$$

where:

k_t = mean conductivity of the liquid-crystal package and plate
δ_t = thickness of the liquid-crystal package and plate

T_ℓ = temperature of surface (liquid-crystal isotherm temperature)
T_b = temperature of water
T_a = temperature of air

The experimental results are presented in terms of a local Nusselt number:

$$N_u = h_\ell \, D_h / k_a \qquad (5)$$

where:

D_h = hydraulic diameter
k_a = conductivity of the air

With the temperature difference between the air and liquid-crystal isotherms fixed, different heat transfer coefficient (Nusselt number) contours are determined by varying heat flux values. The contours of a constant heat transfer coefficient are not directly equivalent to the isotherms, as measured from the photographs or images. They are determined after taking into account thermal conduction in the plate, radiation from the surface, and other corrections, the key one being the lateral conduction. This correction is typically about 4% of the net flux.

The liquid crystal color temperature used is 27.7°C, some 17.6°C below the air temperature (T_a = 45.3°C) for these experiments. Twelve to sixteen isotherms (each corresponding to a different heat flux) are photographed by ordinary 35 mm camera or RGB camera to record the local contours under an oblique Reynolds number. The locations on each isotherm are digitized following a projection of the photographic image onto a digitizing pad or recent color scale representation of the several video images of the heat transfer coefficient distribution into a single image using the Global Lab Color software from Data Translation Inc.

Figure 9 shows the distribution of local Nusselt number for uniform wall temperature on a corrugated-geometry. The contours of constant Nusselt number were obtained using the red color to define a specific isotherm (27.7 for these experiments). Eight to sixteen isotherms (each corresponding to a different heat flux) are photographed and then super-imposed to record local contours under an oblique viewing direction, and then to use human observations. Minimum, maximum, and average Nusselt numbers obtained by numerical integration of this map are also reported. Reproduction of color transparency representing wall temperature contours at Re = 3394 is shown in Figure 10.

Using equations (4) and (5) and the image combination techniques of Global Lab Color features, the color image, Figure 11, was produced. This is a fully standard false color image showing Nusselt number distribution over a central diamond of the corrugated working section.

In Figure 12 the wall temperature contours, due to the square roughness elements at Re = 1.10^4, are shown by (a) image from RGB camera; (b) image through a segmented monochrome line showing the 31.9°C isochrome; and (c) manually processed constant Nusselt number contours. The color image from the RGB camera and the segmentation monochrome line showing the 31.9°C isochrome for the double cylinders arrangement at Re = $2.0 \; 10^4$ are presented together in Figure 13. The image in Figure 14 shows Nusselt number contours around the double cylinders processed by the image combination techniques. By comparing the two presentation methods it can be seen that the process of human interpretation involves simplification of the image. It was thought that comparison to "true" data such as that presented in this study would be more valuable incomputational

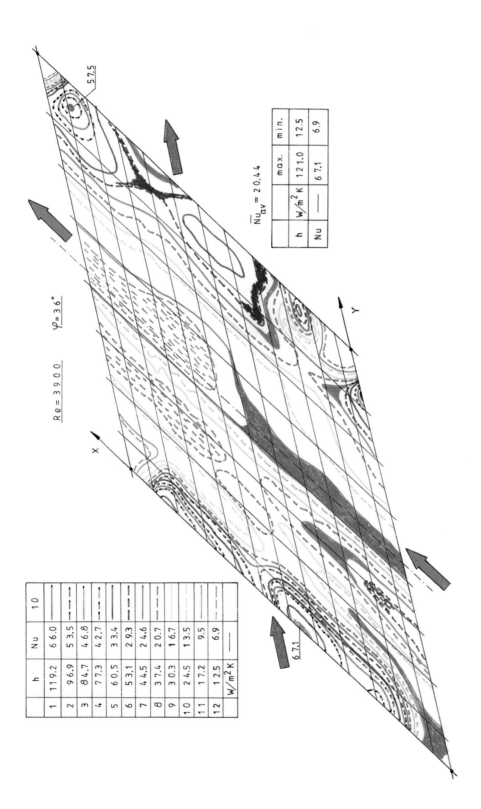

Figure 9 Local heat transfer coefficient and Nusselt number distributions on the bottom wall of the measurement diamond developed from the images in Figure 10.

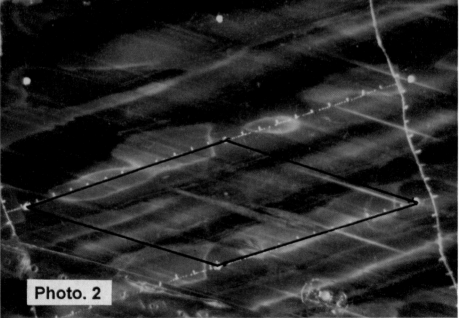

Figure 10   Reproduction of color transparency representing wall temperature contours at Re = 3394 (Photo 1-Exp. 3: Photo 2-Exp. 7; Photo 3-Exp. 9; Photo 4-Exp. 12).

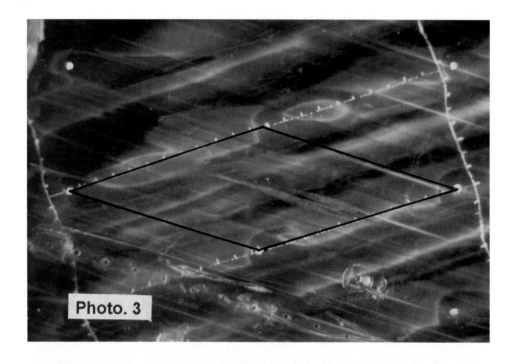

Figure 10 (continued)

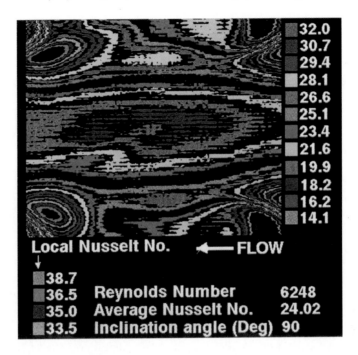

Figure 11   Color scale representation of Nusselt number distribution over a central diamond of the corrugated working section processed by the image combination techniques.

studies because it not only portrays the contour positions, but also gives an indication of the width of the contour, something that is hard to convey using the hand-processed method.

In the second set of experiments (Glycerol-Filled Cavity) the angle of inclination of the cavity could be varied from horizontal through to vertical to observe how the free convection was affected. A temperature gradient of 10K (T_h = 297.6K, T_c = 287.6K) was used so as to obtain desired Rayleigh (Ra = 1.2 10^4), and Prandtl (Pr = 12.5 10^3) numbers and flow conditions in the particular cavity. A sequence of images was taken in the cavity under conditions of steady-state free convection. This sequence was for angles of inclination from horizontal to vertical in 30° steps. A liquid crystal with a color-play range of 3.8K was used, the red color band being 19.6°C. Figures 15 and 16 show images for the midplane of the cavity taking over angles from horizontal to vertical. On the photographs, the liquid crystals convected by the flow appear as a series of colored, uniformly spaced dots. In this particular experiment eight flashes at a time interval of 6 s were used to take one photograph. The displacements of the tracer particles, as recorded on the photographs, enable information to be obtained about the velocity field in the cavity, Hiller et al. (1986).

As mentioned above, in the present experiments we are interested only in the shape of the isotherms but not in their absolute temperature values. For such measurements one needs another calibration procedure, for example, a light source of discrete color spectrum. The isotherms presented in Figure 17 are lines (white) of constant hue range bands for a picture taken under flow resulting from the horizontal configuration of the cavity. This was obtained by taking a photo-image using an RGB video camera and then converting the RGB image to an HSI image. A hue range was selected thus eliminating all other pixels on the image with a different hue value to that range selected. This information may be used to map isothermal contours, as hue has a direct relationship with temperature that is found by calibration.

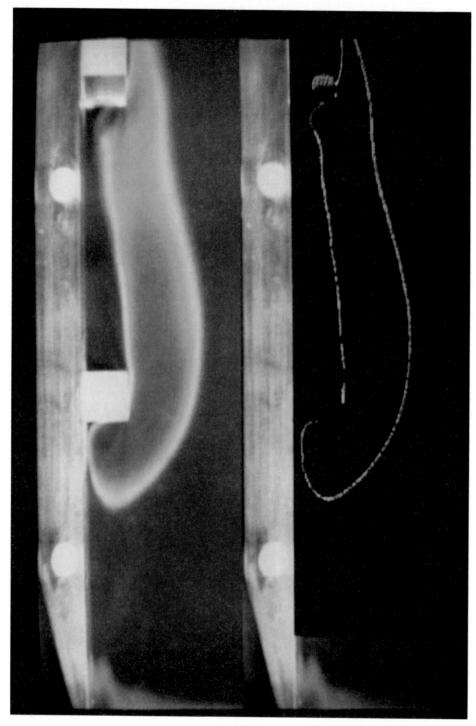

a

b

Figure 12 Heat transfer measurement of the square roughness elements (Re = 1.10⁴; q = 228 W/m²; D_h = 0.009m). (a) Image from RGB camera; (b) segmented monochrome line showing 31.9°C isochrome.

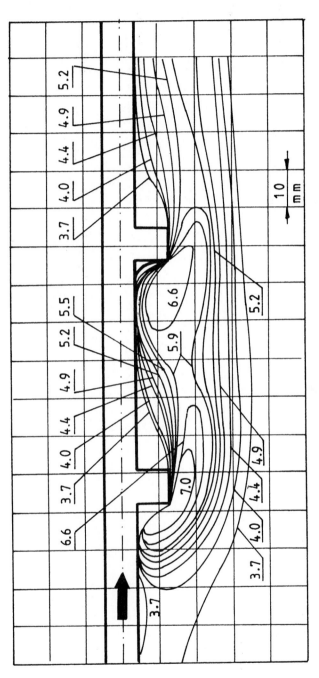

Figure 12 (continued) (c) manually processed Nusselt number contours.

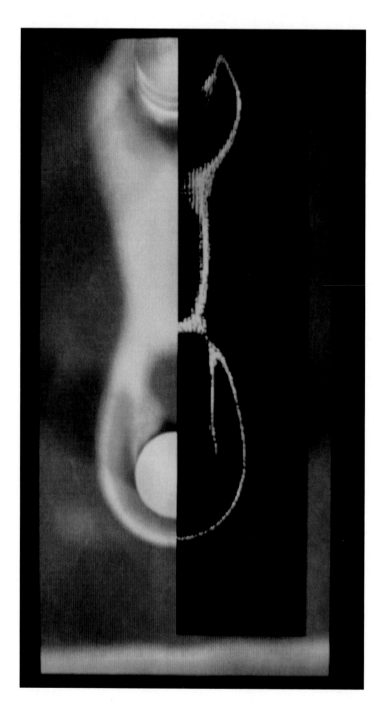

Figure 13   Comparison between the image from the RGB camera and the segmented monochrome 31.9°C line for the double cylinders column at Re = 2.10⁴.

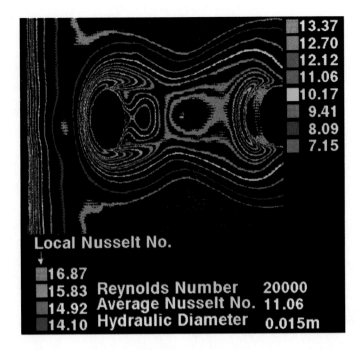

Figure 14   Color scale representation of Nusselt number distribution around double cylinders column processed by image combination techniques.

Conclusions

In this paper we describe briefly the mesophase structure which gives rise to the thermochromic liquid crystal phenomenon, and the application of TLCs to temperature, and also shear stress, measurement. The former is illustrated with typical results from various research investigations of the authors involving forced convection. Stress is laid on the ability of the TLCs to give quantitative data of practical heat transfer significance. Finally, we report the use of TLCs to give both velocity and temperature data in a natural convection cell.

Throughout the paper, we emphasize the desirability and practicality of using automatic processing methods; we have used a *true-color* approach.

Acknowledgment

We are very grateful to PowerGen plc, Radcliffe Technology Centre, UK, for their sponsorship of the cross-corrugated heat transfer project.

Figure 15 Temperature and velocity visualization in a glycerol-filled cavity under free convection using net chiral-nematic liquid crystal (Re = 1.2 10⁴; Pr = 13.5 10³). (a) Horizontal position: 0 = 0° showing Benard cells; (b) vertical position: 0 = 90°.

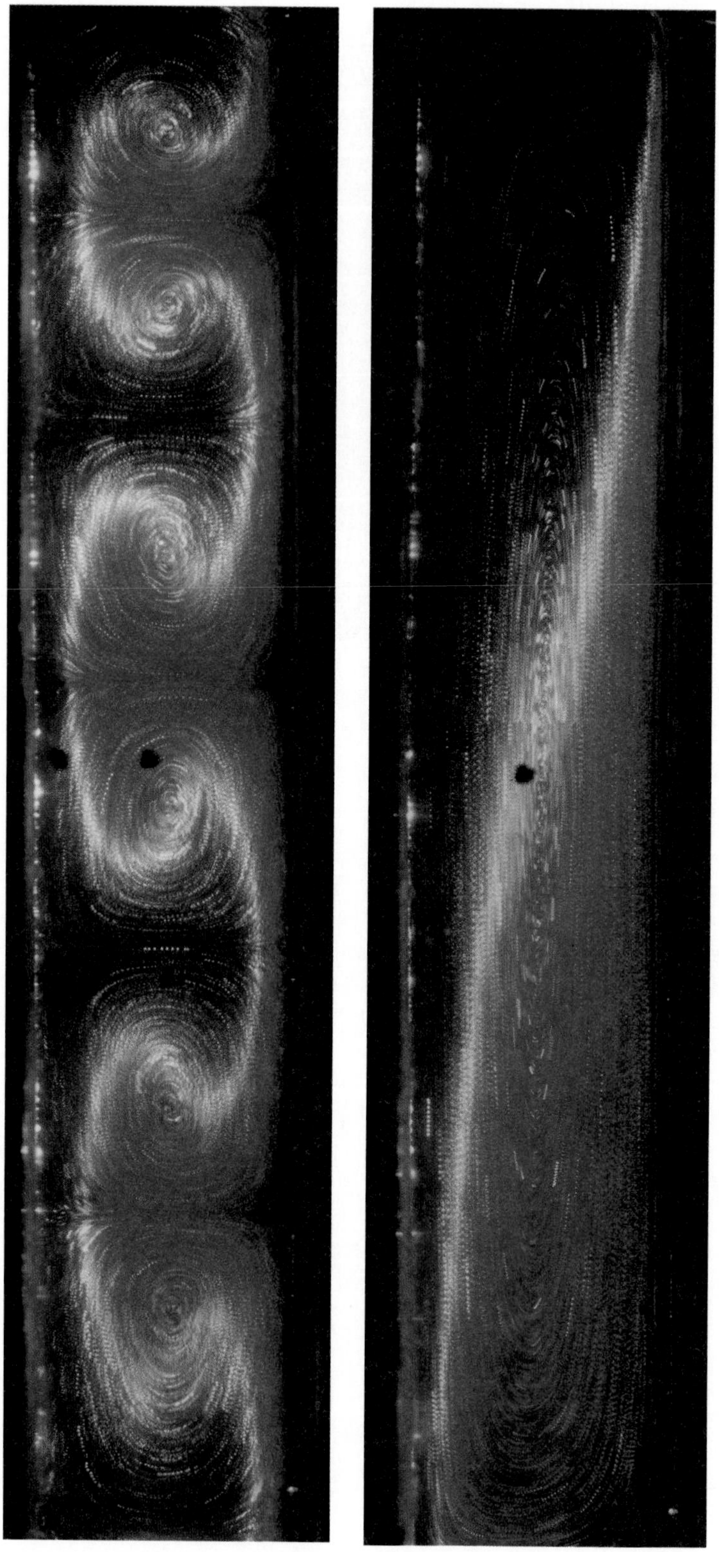

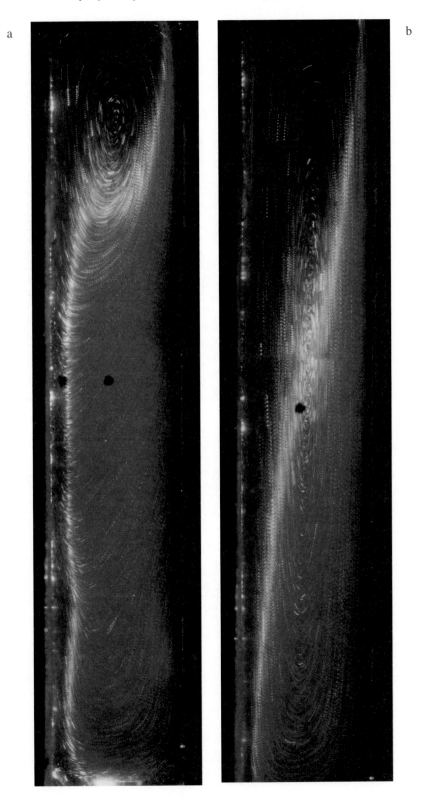

Figure 16 Temperature and velocity visualization in a glycerol-filled cavity under free convection using neat chiral-nematic liquid crystal (Re = 1.2 10⁴; Pr - 12.5 10³). (a) Inclination position: 0 = 30°; (b) inclination position: 0 = 60°.

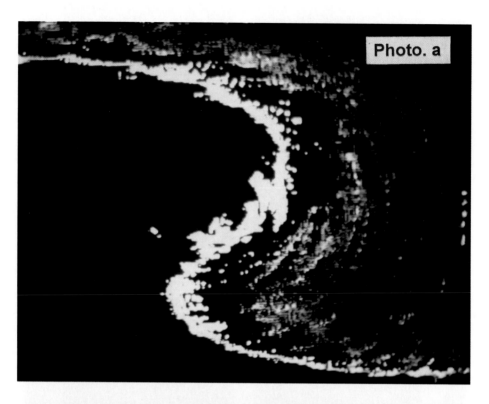

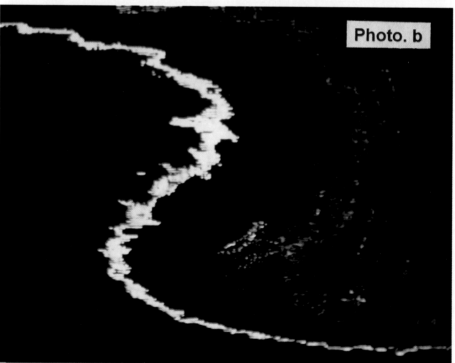

Figure 17 Segmentation transform of the cavity image (one part of Benard cells). (1) Photo (a) - Hue: 40/60; (2) Photo (b) - Hue: 100/120; (3) Photo (c) - Hue: 120/140; (4) Photo (d) - Hue: 140/150.

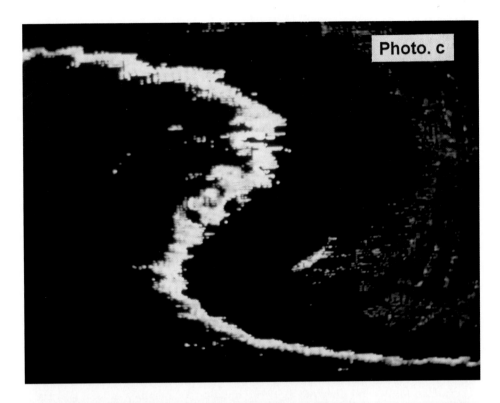

Figure 17 (continued)

References

Akino, N., Kunugi, T., Ichimiya, K., Mitsushiro, K., and Ueda, M., 1989: Improved liquid-crystal thermometry excluding human color sensation. *J. Heat Trans.*, 111, pp. 558–565.

Akino, N., Kunugi, T., Shiina, Y., Ichimiya, K., and Kurosawa, A., 1990: Fundamental study on visualization of temperature fields using thermosensitive liquid-crystals. *Flow Visualization V.* R. Reznicek, Ed., Washington, Hemisphere, pp. 87–92.

Ashforth-Frost, S., Wang, L. S., Jambunathan, K., Graham, D. P., and Rhine, J. M., 1992: Application of image processing to liquid-crystal thermography. *Proc. Opt. Meth. Data Process. Heat Fluid Flow,* City University, London, pp. 121–126.

Bonnett, P., Jones, T. V., and Donnell, D. G., 1989: Shear-stress measurement in aerodynamic testing using cholesteric liquid crystals. *Liq. Crystal,* 6, pp. 271–280.

Baughn, J. W., Ireland, P. T., Jones, T. V., and Saniei, N., 1989: A comparison of the transient and heated-coating methods for the measurements of local heat transfer coefficients on a pin fin. *J. Heat Trans.,* 111, pp. 877–881.

Hiller, W. J. and Kowalewski, T. A., 1986: Simultaneous measurement of temperature and velocity fields in thermal convective flows. *Flow Visualization IV.* C. Veret, Ed., Washington, Hemisphere, pp. 617–622.

Jones, T. V., Wang, Z., and Ireland, P. T., 1992: The use of liquid-crystals in aerodynamic and heat transfer experiments. *Proc. Opt. Meth. Data Process. Heat Fluid Flow,* City University, London, pp. 51–65.

Moffatt, R. J., 1991: Experimental heat transfer. *Proc. 9th Int. Heat Trans. Conf.,* Jerusalem, Vol. 1, pp. 308–310.

Parsley, M., 1988: The use of thermochromic crystals in heat transfer and flow visualisation research. *FLUCOME' 88,* Sheffield University, England, pp. 216–220.

Reinitzer, F., 1888: Beiträge zur Kenntniss des Cholestrins, *Monatschr. Chem. Wein,* 9, pp. 50–90.

Stasiek, J., Collins, M. W., and Chew, P. E., 1991: Liquid-crystal mapping of local heat transfer in crossed-corrugated geometrical elements for air heat exchangers. *EUROTECH DIRECT '91 Congress,* Birmingham, England, Paper C413/040.

Stasiek, J. and Collins, M. W., 1989–92: *Internal Reports,* City University, London.

Stasiek, J. and Collins, M. W., 1992: Liquid-crystal thermography and image processing in heat and fluid flow experiments. *Flow Visualization VI.* Y. Tanida and H. Miyashiro, Eds., Springer-Verlag, 1992, pp. 439–450.

B.D.H. Limited. Advanced Materials Division, Broom Road, Poole, Dorset BH24 NN, United Kingdom.

DATA TRANSLATION, 1991: *Image Processing Handbook.*

chapter six

The Applications of the Hydrogen Bubble Method in the Investigations of Complex Flows

Qi Xiang Lian[1] and Tsung-chow Su[2]

[1]Institute of Fluid Mechanics, Beijing University of Aeronautics and Astronautics, Beijing, China
[2]Department of Mechanical Engineering, Florida Atlantic University, Boca Raton, Florida, USA

Abstract — Some typical results of investigations of complex flows with large vortical structures using the hydrogen bubble method are discussed. Comments are made about the behavior of the hydrogen bubble marks, the capability and limitation of this method, and the implications of the hydrogen bubble-visualized pictures.

Introduction

Although there is no clear definition for the complex flow, generally the separated flows, turbulent flows, and multivortices flows are considered complex flows. The majority of the flows in industries or in nature might be complex flows. Complex flows are generally three-dimensional (3-D), unsteady, and have multivortices. An ideal experimental method for the investigation of complex flows should be able to depict the instantaneous 3-D velocity field. Particle image velocimetry (PIV) and the speckle method might be useful but still have great difficulties in the measurement or visualization of the instantaneous 3-D flow field. The hydrogen bubble method is relatively simple, inexpensive, and convenient; it might still be a useful means for the investigation of certain complex flows, especially for basic researches. In this chapter, some significant results obtained in the investigations of some complex flows using the hydrogen bubble method are presented, and discussions are made about the behavior of the hydrogen bubble marks, their limitations, and capability for depicting a complex flow structure.

The Applications of the Hydrogen Bubble Method in the Investigation of Coherent Structures of Turbulent Boundary Layers

The coherent structures of turbulent boundary layers are some of the most complicated flows in fluid mechanics. In the 1960s, Kline et al. (1967) applied the hydrogen bubble method to the study of the near-wall flows in the turbulent boundary layer and discovered the low-speed streaks, high-speed streaks, and the bursting phenomenon. Later, Kim et al. (1971), discovered vortical flow structures in turbulent boundary layers using the hydrogen bubble method. Thereafter many authors used this method in the study of turbulent boundary layers, such as Nakagawa and Nezu (1981), Smith and Metzler (1983), Lian (1983, 1985, 1990), and Talmon et al. (1986).

Hydrogen bubbles as visual tracers are far smaller than the scales of the coherent structures studied. The interference may generally be neglected, and the characteristic time of the bubbles is usually in the order of 10^{-5} s. Thereafter the bubbles can generally follow the instantaneous flow speed of the fluid, thus they may reflect the features of the unsteady coherent structures. The hydrogen bubble method is the most widely used technique in the visual study of coherent structures, many important results are obtained from it.

From investigations in the past decades, it is generally acknowledged that the vortical structures in the turbulent boundary layers play an important role. There were attempts to synthesize the flow in a turbulent boundary layer by a certain combination of vortices (such as Λ vortices). Correct modeling of the flow by vortices requires knowledge of the vortical structures, such as their shapes, their locations relative to high/low-speed streaks, and their developing and vanishing processes. The measurement of even a single unsteady, 3-D vortex in a turbulent boundary layer by various velocimetry methods is still very difficult, and there is very limited literature about these vortical structures by visualization other than by use of the hydrogen bubble technique. In Lian's paper (1990), the streamwise, or xy, vortices were observed, and their locations were always at the interfaces between the high/low-speed streaks. The transverse vortices were always located at the front interface of the high-speed streaks where the upstream is the high-speed region and the downstream is the low-speed region. In Figure 1 the plan views show the shapes and the locations of the streamwise vortices and transverse vortices.

The rise of low-speed streaks had been observed using dye injection by Kline et al. (1967). In Lian's paper (1990), the plan view was illuminated by an ordinary lamp and an oblique laser sheet simultaneously. A pair of photos at normal exposure and long exposure could show the rise of the low-speed streaks and the vortical rotation of the rising fluid. Figure 2 is an example of long exposure showing the rise of fluid from the wall. The average inclination of the rising fluid in low-speed streaks could be measured from the laser sheet cut sections, and also could be shown by side views (Lian 1990). The inclination of the leg of the horseshoe vortex was predicted by Theodorsen (1952) to be 45°, the side views showed that a large amount of the rising fluid might be included about 45° as shown in Lian's Figure 14 (1990).

The Investigations of Vortices in the Front Stagnation Region

The investigation of the intensification of random eddies or turbulence in the stagnation region was carried out early in the 1960s by Sutera (1965) and by Sadeh et al. (1980). As pointed out by Sadeh et al., the amplification of the turbulence is due to the stretching of the vortex lines as they approach the stagnation region. In recent years the present authors, in their investigation of disturbed flows in super clean rooms, discovered that very small vortices or vorticities generated by a fine wire in front of a bluff body might be organized into a very large vortex (Su and Lian 1989).

Figure 3 shows the flow pattern in front of a square plate while the oncoming flow is uniform. Figure 4 shows the large vortex pair formed in the front stagnation region when there is a tiny wire placed in the upstream. The wire is placed in the plan of symmetry of the square plate. The wire's distances to the plate were varied from 0.5 to 2 B, B being the width of the plate. The wire's diameter was varied from .1 to 1 mm. The Reynolds numbers of the wire might be changed from below 50 to more than 100, the wake of the wire might

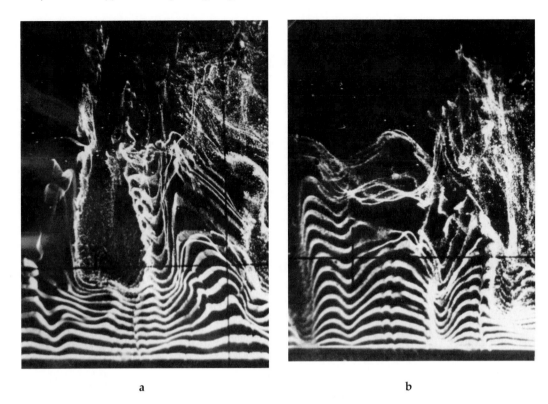

a   b

Figure 1   A plan view shows the streamwise and the transverse vortices in a turbulent boundary layer at y+ = 11: (a) streamwise vortex; (b) transverse vortex (Lian, 1990).

either be a laminar wake or be a vortical flow with vortex street. In all these cases, large vortices could generally be formed in the front stagnation region.

In another test, the upper edge of the plate is above the water surface and the interference wire is just touching the water surface, therefore the stagnation region is in front of the plate. Figure 5 shows the large vortex formed while there is an interference wire in the upstream. There is no counter-rotating vortex near this vortex, as in the case when the plate is totally immersed in the water as in Figure 4. Therefore there is no loss of vorticity due to the vortex of opposite rotation, so this vortex is larger and exists continuously. While in the case of Figure 4, the vortex pair is not stable and forms and breaks repeatedly.

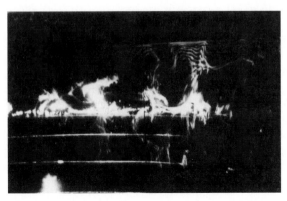

Figure 2   The laser sheet oblique to a plan view showing the rise of the fluid in the low-speed streaks.

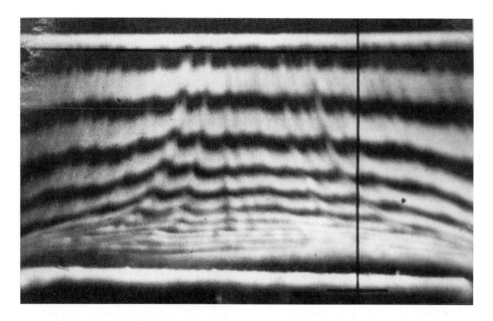

Figure 3   The flow pattern in front of a square plate without a disturbing wire.

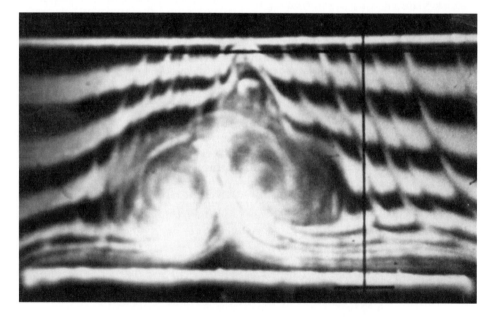

Figure 4   The large vortex in front of a square plat with a disturbing wire.

The large vortex formed due to a tiny interference wire might be hundreds of times larger than the wire. This phenomenon might be interesting for theoretical investigations and also may have certain significance for various applications. Further experimental results and discussion about the mechanism of the formation of large vortices are in Lian and Su (1994).

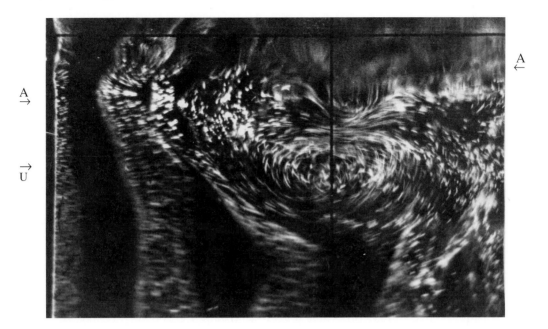

Figure 5   The large vortex in front of a square plate at the water's surface with a disturbing wire. The arrow A points to the water surface.

Discussions of the Hydrogen Bubble Visualized Pictures

The previous sections present some examples of the applications of the hydrogen bubble method in the study of complex flows and vortices. The hydrogen bubble-visualized pictures depend on the behavior and the features of the hydrogen bubble itself, yet there is very limited literature on this subject. In the following sections some of the features which influenced the visualized pictures are discussed.

The Life of Hydrogen Bubbles

After being released from the generating wire, hydrogen bubbles gradually dissolve in water. In a uniform flow one could see that the hydrogen bubble time lines vanish in the downstream direction, which appears as the decrease of their brightness. At a certain distance downstream of the generating wire, all hydrogen bubble time lines vanish. The vanishing course of the hydrogen bubbles is influenced by many factors, but in a uniform flow, almost every time line disappears at nearly the same distance from the platinum wire; this means the life span is almost the same. In the turbulent boundary layer, or in other kinds of complex flows, the situations are quite different. In certain regions the lifetime of the hydrogen bubbles is shortened, and in other regions it might be lengthened. The life of the hydrogen bubbles is related to the local structure of the flow. The lifespan of the hydrogen bubbles in the turbulent boundary layer was discussed briefly by Lian (1990), in the present chapter it will be discussed in detail.

First, let us consider a small hydrogen bubble in water. Let the relative movement of the bubble to the water be neglected. This assumption is reasonable, for example, the terminal velocity due to buoyancy of a bubble in liquid might be calculated under the balance of the buoyancy force and the drag force. The drag force of a gas bubble in liquid is 4 π μ ur (Lamb, 1932). Thus the terminal velocity of the rising gas bubble is:

$$u = gd^2 (\rho - \rho_h)/12\mu \qquad (1)$$

where d is the bubble diameter and ρ is the density. In some literature, the Stokes formula was used to calculate the drag force of the bubbles. But the nonslip boundary condition required by the Stokes formula is valid for solid particles, not for gas bubbles. Therefore Equation (1) might be better, though the difference is not large. At room temperature, the upward floating velocity of a bubble of .01 mm in diameter is about .008 cm/s. The relative velocity of the bubble in water due to local flow acceleration is generally much smaller than this value, because the flow acceleration is generally much smaller than gravitational acceleration, g. Neglecting the relative motion, and assuming the concentration of hydrogen C is distributed symmetrically around the bubble, the dissolving rate of hydrogen in the bubble is:

$$dm/dt = 4\pi r^2 k (Cs - C) \qquad (2)$$

where C is the concentration of hydrogen in water around the bubble, Cs is the saturation concentration of hydrogen in water, and k is a proportional constant, which varies with temperature and pressure and thus might be considered a constant. If the distances between the bubbles are large and the distribution of C reaches an equilibrium, C might be considered constant. The density of hydrogen gas inside the bubble might also be considered as constant. Thus from the above equation the following equation might be derived:

$$dr/dt = -k (Cs - C)/\rho_h \qquad (3)$$

where ρ_h is the density of the hydrogen bubble. Assuming C is nearly constant, then the diameter decreases almost linearly and the life of a bubble is proportional to its diameter. Equation (3) may be used for the correlation of the life of hydrogen bubbles with certain characteristics of the flow structure. If the local fluid flow tends to concentrate the hydrogen bubbles, the life of the hydrogen bubbles will be lengthened. It is very difficult to study the features of the hydrogen bubbles in detail, but it is possible to discuss some extreme, global behaviors of the bubbles in some regions.

The concentration C of the hydrogen in the water around a bubble is difficult to determine. However, here Ca is introduced as the regional average concentration, which is the total mass of the hydrogen in the bubbles in a small unit volume at a time soon after the formation of the time lines. Ca would be uniformly distributed in a uniform flow. When the bubbles are concentrated to the vortex core, the local Ca is increased. When the fluid with bubble time lines is stretched, the local Ca is decreased. In an accelerating flow the bubbles would move relative to the fluid, just like the floating due to buoyancy in a gravitational field, and the relative velocity of the bubble to the fluid might be determined by Equation (1), only replace g by the local flow acceleration a. By equating the local buoyancy force due to acceleration and drag force, and neglecting the inertia force, the characteristic time is very small, on the order of 10^{-5} s. The equilibrium of the buoyancy force and the drag force on a bubble will be reached in a very short time, on the order of 10^{-4} s.

The Inflow of the Hydrogen Bubbles to the Core of a Vortex

Consider a 2-D vortex, assuming the vorticity distribution is a Rankine model, inside the core of the vortex the rotational velocity is a constant ω. Using the above argument and equations, letting r denote the distance of a bubble to the core center, it is easy to derive:

$$r/r_i = \exp\,(-Re_d \omega t/12) \tag{4}$$

where r_i is the position of a bubble at t = 0, and r is its position at time t; $Re_d = \omega d^2/v$. At a characteristic time $t^* = 12\,\ln(2)/Re_d\omega$, every bubble moves to a position at $r_i/2$, which implies that the distances between any two bubbles decreases to half of its initial value while the average concentration of the hydrogen bubbles Ca increases 4 times. The characteristic time $t^* = 12\,Ln(2)/Re_d\omega = .083/\omega^2 d^2$. Obviously the larger bubbles are moving much faster towards the center of the vortex core. The small vortices of a starting vortex may have a large rotational speed as shown by Lian (1989), the ω of the core may be on the order of 50/s, for bubbles of 0.0025 cm the characteristic time is about 1.6 s. Actually the rotational speed of a vortex core may be much larger and the motion of the hydrogen bubbles toward the core of a vortex may be very remarkable, as shown in Figure 6. In Figure 6, the flow is the same as Figure 5. There is a large vortex in the front stagnation region. This photo was taken with the camera oblique to the water surface so that the shape of the vortex core could be observed. In motion picture films, the flow of the hydrogen bubbles towards the core could be observed. In Figure 6 there are two frames showing the concentration of the hydrogen bubbles at the core and also showing the axial flow of the bubbles in the core. The centrifugal acceleration causes the hydrogen bubbles to concentrate at the core as indicated by Equation (4). Also, the axial flow in the core causes the surrounding fluid flowing to the core, which carries bubbles, to further increase the concentration of hydrogen bubbles. Therefore, the core shows a high density of hydrogen bubbles (see Figure 6b). This is the reason, after hydrogen bubbles have stopped being released, the figure of the vortex remains clear while the figures of other structures vanish, as discussed by Lian (1990).

The Black Spot in the Plan View of the Turbulent Boundary Layer

The other extreme is the *black spots* which generally appear in the high-speed region of the near wall in plan views of turbulent boundary layers. The hydrogen bubble time line vanishes much earlier than those surrounding it in a black spot, thus it leaves a black region among the white hydrogen bubble lines. In Figure 7 the arrow marks an example of the black spot. Talmon et al. (1986) considered the black spot as the footprint of the fluid ejection from the wall region. However, motion picture films show that inside a black spot the fluid is generally stretching in an xz plan. In Figure 7, one can see that inside the black spot the spacings of the hydrogen bubble time lines is greater than those around it. From Figure 7a to 7b, the hydrogen bubble time lines are diluting, the spacing between them is widening, and the black region is enlarging. These phenomena imply that the fluid inside might be stretching. From the equation of continuity, if

$$\delta u/\delta x + \delta w/\delta z > 0 \tag{5}$$

then $\delta v/\delta y < 0$. Since v = 0 at y = 0, then v = 0 near the wall. This is a flow approaching the wall. It is the feature of a "sweep flow". Due to continuity, the inward flow over the black spot should be simultaneously balanced by outward flow. The laser sheet-illuminated plan view shows that, generally, there is an intense rise of the fluid in the low-speed streaks near the black spots.

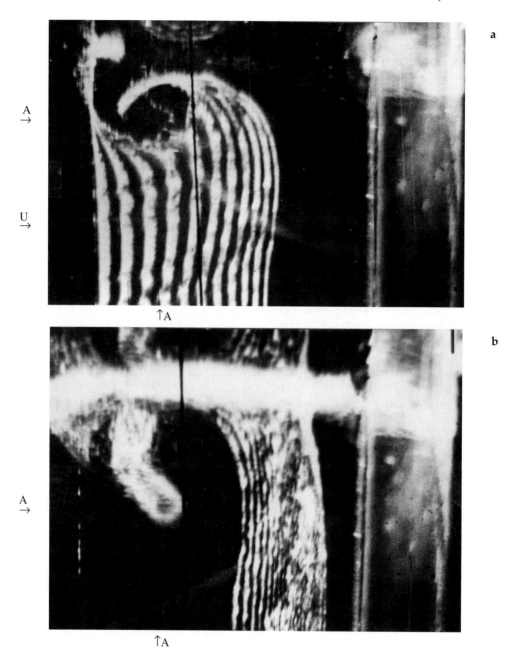

Figure 6 The concentration of hydrogen bubbles at the core of a vortex. The hydrogen bubbles were released only for a short time period: (a) the photo shows the large vortex; (b) 2.87 s later, the photo shows the hydrogen bubbles flowing to the core and the axial flow in the core. The arrow A points the core of the vortex.

The Effect of Turbulence on the Appearance of the Time Lines

In the region where the scale of turbulence is smaller than the width of the bubble lines, the fluid with hydrogen bubbles mixes quickly with surrounding fluid, thus the hydrogen bubbles dissolve more quickly. This was observed by the senior author in the investigation of the large vortical structures in the reattaching flow of a laminar boundary layer behind a backward step (Lian 1993, *ACTA Mechanica SINICA*, vol. 25, pp. 129–133). Figure 8 shows

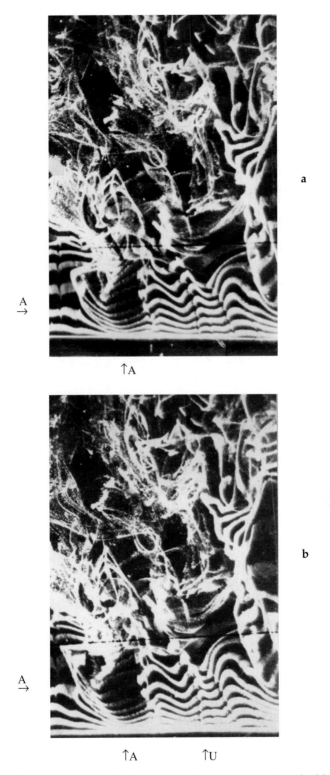

Figure 7 An example of the picture of a black spot, the arrow A points the black spot: (a) t = 0; (b) t = .067 s.

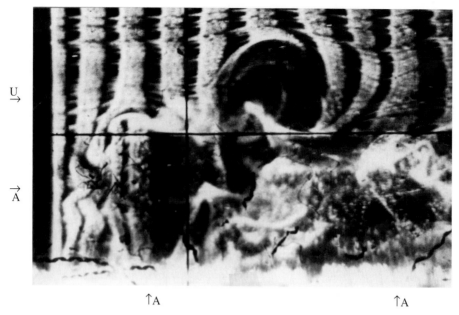

Figure 8 An example of the turbulent region near a large vortex in a backward step flow, the oncoming boundary layer before separation is laminar. The arrow A points to the region where the time lines have inflection points, the flow becomes turbulent, and the hydrogen bubbles vanish quickly.

an example (the arrow points a black region) of where hydrogen bubble lines diffuse quickly and their edges become chaotic. In comparison, the bubble lines in the other region are smooth with sharp edges.

Conclusion

The hydrogen bubble technique is useful in the investigation of complex flows, especially for flows with vortical structures, as the hydrogen bubbles tend to concentrate at the cores of the vortices.

Acknowledgments

The senior author wishes to express thanks for the financial support of the Chinese Natural Science Foundation and the Chinese Aeronautical Foundation. The second author acknowledges the support of the Office of Naval Research under Grant Number N00014-91-J-1420.

References

Kim, H. T., Kline, S. J., and Reynolds, W. C.: The production of turbulence near a smooth wall in a turbulent boundary layer, *J. Fluid Mech.*, Vol. 50, pp. 133–160, (1971).

Kline, S. J., Reynolds, W. C., Schraub, F. A., and Rundstadler, P. W.: The structures of turbulent boundary layer, *J. Fluid Mech.*, Vol. 30, pp. 741–773, (1967).

Lamb, H.: *Hydrodynamics* (6th Ed.), Cambridge University Press, p. 602, (1932).

Lian, Q. X.: The observation of coherent structures in turbulent boundary layer in flows with adverse pressure gradient, *Acta Mech. SINICA*, Vol. 4, pp. 414–418, (1983).

Lian, Q. X.: The structures of turbulent boundary layer in the vicinity of separation, *Acta Mech. SINICA*, Vol. 1, pp. 71–80, (1985, English edition).

Lian, Q. X.: A visual study of the coherent structure of the turbulent boundary layer in flow with adverse pressure gradient, *J. Fluid Mech.*, Vol. 215, pp. 101–124, (1990).

Lian, Q. X. and Huang, Z.: Starting flow and structures of starting vortex behind bluff bodies with sharp edges, *Exp. Fluids*, Vol. 8, pp. 95–103, (1989).

Lian, Q. X. and Su, T. C.: Large vortex in front stagnation region of a square plate induced by a fine wire, *Science in China (Series A)*, Vol. 37, pp. 469–477, (1994).

Nakagawa, H. and Nezu, I.: Structures of space-time correlations of bursting phenomena in an open channel, *J. Fluid Mech.*, Vol. 104, pp. 1–43, (1981).

Sadeh, W. Z. and Brauer, H. J.: A visual investigation of turbulence in stagnation flow about a circular cylinder, *J. Fluid Mech.*, Vol. 99, pp. 53–64, (1980).

Smith, R. C. and Metzler, S. P.: The characteristics of low-speed streaks in the near wall region of a turbulent boundary layer, *J. Fluid Mech.*, Vol. 129, pp. 27–54, (1983).

Su, T. C. and Lian, Q. X.: On clean room fluid dynamics. MS 89-338, Conference on Automatic Clean Room Processes *(Am. Soc. Manuf. Eng.)*, March 21–22, Kissimmee, Florida, (1989).

Sutera, S. P.: Vorticity amplification in stagnation point flow and its effect on heat transfer, *J. Fluid Mech.*, Vol. 21, pp. 513–534, (1965).

Talmon, A. M., Kunen, J. M. G., and Ooms, G.: Simultaneous flow visualization and Reynolds-stress measurement in a turbulent boundary layer, *J. Fluid Mech.*, Vol. 163, pp. 456–478, (1986).

Theodorsen, T.: Mechanism of turbulence, *Proc. 2nd Midwestern Conf. Fluid Mech.*, Ohio State University, Columbus, Ohio, (1952).

chapter seven

Surface Flow Visualization of a Backward-Facing Step Flow

Giovanni Maria Carlomagno

University of Naples, DETEC, P.le Tecchio, 80125 Naples, Italy

Abstract — *Surface flow visualization and heat transfer measurements in a backward-facing step flow are performed by means of an infrared scanning radiometer on both sides of the channel downstream of the step. In the range 260 < Re < 500 the flow is substantially two-dimensional (2-D) with only one separated zone whereas, regardless of the high channel aspect ratio, a three-dimensional (3-D) flow is found for 500 < Re < 5000. An additional region of flow separation, downstream of the step and on the step side of channel, is detected in the range 1400 < Re < 3400. An unstable flow configuration is also encountered for 1400 < Re < 1950. In the range 5000 < Re < 20000 the flow is again substantially 2-D with one separated region on each side of the channel. In the vicinity of the wall opposite to the step and of the step itself, the presence of one or more vortexes, which seem to rotate about an axis normal to the two wide walls of the channel, is found for 20000 < Re < 39000. For Reynolds number further increasing a pronounced waviness of the color bands of the thermograms is still present which practically vanish for Re = 42000. In the range 42000 < Re < 50400 a 2-D turbulent flow is recovered.*

Introduction

When a duct flow encounters an abruptly enlarged area, it separates from the wall and, depending on the downstream conditions, the separated flow will eventually reattach and undergo a redevelopment. Separated flows which reattach may cause a large variation of the local heat transfer coefficient values. The nonsymmetric abrupt enlargement in a parallel-plate channel, generally referred to as backward-facing step, has perhaps been investigated more than any other separated flow situation [Abbot and Kline (1962), Seban (1964), Sparrow et al. (1987)].

The most comprehensive recent fluid dynamic investigation (in the laminar, transitional, and early turbulent flow regimes) on the backward-facing step flow has been probably carried out by Armaly et al. (1983) in the Reynolds number range *70 < Re < 8000*. Beyond the expected primary zone of recirculating flow attached to the step corner, they demonstrate that additional regions of flow separation, downstream of the step and on both sides of the channel, are present in the transitional regime. Furthermore they show that the flow downstream of the step only remains 2-D at low and high Reynolds numbers (*Re < 400* and *Re > 6000*).

Heat transfer experimental investigations can be separated into three groups: laminar [Aung (1983), Sparrow et al. (1987), Cardone et al. (1993)], transitional [Aung and Goldstein (1970), Kottke (1983), Cardone et al. (1993)] and turbulent [Seban (1964), Vogel and Eaton (1985), Scherer et al. (1993)] The streamwise distribution of the convective heat transfer coefficient, downstream of the step and on the step side, is generally characterized by an initial increase that leads to the attainment of a maximum, which is followed by a subsequent decrease. Further downstream the distribution recovers the typical behavior of the redeveloped flow. No heat transfer data relative to the wall opposite to the step seems to have been published by other authors so far.

The aim of this work is to deepen the understanding of the backward-facing step flow by performing surface flow visualization as well as heat transfer measurements on both sides of the step by means of an infrared scanning radiometer (IRSR). Application of IRSR to this problem is advantageous on account of its relatively good spatial resolution and thermal sensitivity. Moreover, the use of IRSR matches both qualitative and quantitative requirements. The essential features of the radiometer are: it is nonintrusive, it allows a complete 2-D mapping of the surface to be tested, the video signal output may be treated by digital image processing [Carlomagno and de Luca (1989), de Luca et al. (1990)].

Experimental Apparatus and Procedure

The open air-driven flow channel used for this study is shown schematically in Figure 1. The 2-D backward-facing step provides an expansion ratio of 1:2. The larger channel, downstream of this step, has a height of 10 mm and an aspect ratio of 17:1. The tunnel and the test section are constructed mainly from aluminum and all parts are machined to very close tolerances with regard to parallelism of walls, surface roughness, and manufacturing of step corners. The air flow originates from a large settling chamber, having flow straighteners and screens, and is afterwards guided into a converging nozzle whose walls have a 30 mm radius of curvature. The nozzle outlet is connected to the inlet channel, which is 5 mm in height and 200 mm in length up to the step. These dimensions ensure a 2-D almost fully developed flow at the cross-section where the step is located. The two wide walls of the channel after the step, for a length of 200 mm, are made with a very thin costantan foil (50 µm in thickness) that is heated by Joule effect; the foil is embedded in a 50 µm groove cut in a bakelite frame which in turn is flush mounted with the aluminum walls. Each foil is kept taut by means of springs which act on two couples of copper clamps; a thin indium

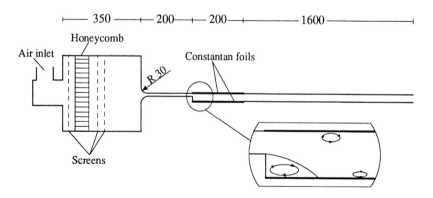

Figure 1   Air tunnel and test section.

wire is inserted in between foil and clamps to obtain a good electric contact. In order to raise the surface emissivity coefficient ε of the viewed surface, the foil is coated there with a thin layer of a black paint which gives $\varepsilon = 0.95$ in the wavelengths of interest.

An IR camera takes alternating temperature maps of both sides of the channel downstream of the step; these maps are correlated to the heat transfer coefficient by means of the *heated-thin-foil* technique used by Carlomagno and de Luca (1989). In particular, for each pixel, the convective heat transfer coefficient h is calculated as:

$$h = (q_w - q_r)/(T_w - T_b) \tag{1}$$

where q_w is the Joule heat flux, q_r the radiative flux to ambient, T_w and T_b are the wall temperature and the bulk temperature, respectively. To insure relatively high temperature differences, tests are performed with q_w values ranging from 450 to 4400 W/m^2 according to the air flow rate. The bulk temperature is evaluated by measuring the stagnation temperature just before the step and by making a 1D energy balance along the channel. Heat transfer coefficients are shown in nondimensional form by means of the local Nusselt number:

$$Nu = h/D/\lambda \tag{2}$$

where D is the hydraulic diameter of the *inlet* channel and λ the thermal conductivity coefficient of air evaluated at film temperature.

Tests are carried out for Reynolds number Re ranging from 260 to 50400. Re is defined in the conventional way:

$$Re = VD/\nu \tag{3}$$

where V and ν are the average velocity in the channel before the step and the kinematic viscosity coefficient of air.

The regions of separation strongly influence the value of the convective heat transfer coefficient h; in particular the streamwise distribution of the heat transfer coefficient within and downstream of the separated region on the step side is typically characterized by an initial increase of h which leads to the attainment of a maximum and a subsequent decrease. Further downstream, the distribution takes on the trendwise characteristics that correspond to the redeveloped flow. In the following, the aforementioned heat transfer maximum, which occurs at a streamwise location x_{max}, is assumed to coincide with the location x_r at which the flow reattaches. There is, however, considerable evidence that the assumed equality of x_{max} and x_r is by no means universal as shown by Sparrow et al. (1987).

The employed IR thermographic system is based on AGEMA Thermovision 900. The field of view (which depends on the optics focal length and on the viewing distance) is scanned by the Hg-Cd-Te detector in the 8-12 μm IR band. Nominal sensitivity, expressed in terms of noise equivalent temperature difference, is 0.07°C when the scanned object is at ambient temperature. The scanner spatial resolution is 230 instantaneous fields of view per line at 50% slit response function. A 5° × 10° lens is used during the tests at two viewing distances of about 1.2 m and 2 m which give a field of view of 0.2 × 0.1 m^2 and 0.34 × 0.17 m^2 respectively. An application software has been developed that can perform noise reduction by numerical filtering, computation of temperatures, and heat transfer correlations on the thermal image. The latter is digitized in a frame of 272 × 136 pixels × 12 bit. At this stage, no attempt to correct errors due to tangential conduction within the foil has been made; it is believed, however, that these errors are within few percent of the measured values for most of the measurements.

Results

In the range $260 < Re < 500$ the flow appears essentially 2-D. However, regardless of the high aspect ratio of the channel and in substantial agreement with Armaly et al. (1983), temperature maps show a 3-D idea of the flow downstream of the step in the Reynolds number range $500 < Re < 5000$; in any case the flow practically maintains its symmetry to the centerplane of the test section. This behavior is shown in the thermograms of Figures 2 and 3 where temperature maps on both sides of the channel downstream of the step are represented for $Re = 480$ and 1000 respectively. In these thermograms the location of x_{max} is in the neighborhood of the regions where the temperature attains minimum values. Flow is from left to right and the step is practically located at the left margin of each thermogram. As it may be noticed, aside from edge effects, the flow at $Re = 480$ is substantially 2-D while at $Re = 1000$ a 3-D regime appears. Figures 2a and 3a show also that for increasing Reynolds number the reattachment zone, which is in the vicinity of the zone of minimum temperature, moves downstream.

At higher Reynolds number values, and in particular for $1400 < Re < 1950$, two different flow configurations are encountered. The first one, referred in the following with the letter U, is an *unstable* flow configuration; this configuration is only reached by slowly increasing the flow rate that enters the test section. In this regime, any substantial disturbance applied to the flow and/or to the channel walls leads to the second flow configuration, *stable*, which is referred to in the following by the letter S. The second flow configuration is also attained when, starting from a relatively high flow rate, this latter is slowly reduced. The thermogram of Figure 4a, which is relative to the stable flow configuration

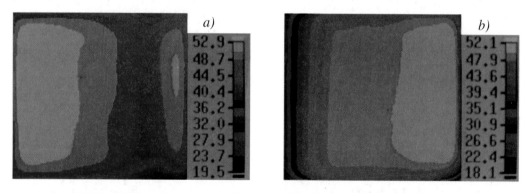

Figure 2   Temperature maps for Re = 480: a) lower wall; b) upper wall.

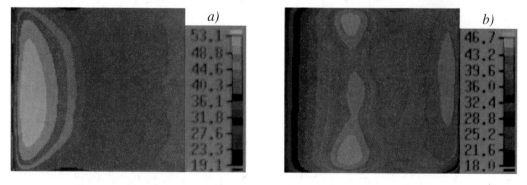

Figure 3   Temperature maps for Re = 1000: a) lower wall; b) upper wall.

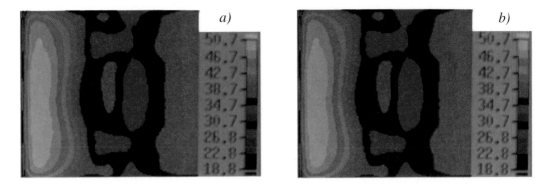

Figure 4   Temperature maps on lower wall for Re = 1400: a) stable; b) unstable.

for $Re = 1400$, shows that, moving along the centerline of the channel from left to right, the temperature first decreases attaining a minimum in the first dark-brown zone and then increases (orange zone); afterwards the temperature decreases again in the green zone and finally increases continuously to the right end of the thermogram. The two temperature minima correspondingly indicate two maximum values of the convective heat transfer coefficient [see Equation (1)] and therefore two reattachment zones. The other two green zones, symmetrically located with respect to the centerline of the channel in the middle of the thermogram, also indicate high values of h. The situation depicted by the thermogram of Figure 4b, which is relative to the unstable flow configuration for the same Re, is completely different; in this case only one minimum temperature zone appears (the green one) so that only one reattachment zone in present. Both thermograms of Figure 4 are relative to the lower wall downstream of the step. For the tested apparatus, the unstable flow configuration does not appear for $Re > 1950$.

The three-dimensionality of the flow is kept from $Re = 500$ up to about $Re = 5000$. Additional regions of flow separation are found in the range of $1400 < Re < 3400$ on the step side of channel. In agreement with Armaly et al. (1983) the flow appears to maintain its two-dimensionality only immediately downstream of the step.

Figures 5a and 5b, which are relative to $Re = 7500$, show the reestablishment of a 2-D flow. It has to be stressed however that a certain degree of regular waviness in the color contours appears, especially at the lower wall.

As shown in Figure 6, where Nusselt number profiles on the centerline of the lower wall are reported for several Re, the laminar regime of the flow is characterized by a

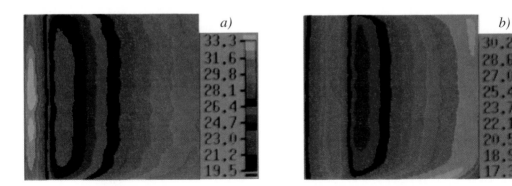

Figure 5   Temperature maps for Re = 7500: a) lower wall; b) upper wall.

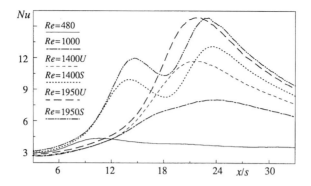

Figure 6   Nusselt number profiles on the centerline of the lower wall.

reattachment length that increases with increasing Reynolds number. In Figure 6, and in the following figures, the distance x from the step location is nondimensionalized with respect to the step height s which, in turn, is equal to the inlet channel height.

At higher Reynolds number values, and in particular for $Re = 1400$ and $Re = 1950$, two different Nu profiles are present according to the flow configuration. The *unstable* flow configuration exhibits only one peak of heat transfer coefficient while the *stable* configuration is characterized by the presence of two peak values of Nu, the first peak being lower than the second one.

For further increasing Re (see Figure 7), only the second flow configuration is obtained, the two peaks move upstream and the first peak increases more rapidly than the second one so that the latter is practically swallowed by the tail of the first peak at about $Re = 3400$.

The presence of the two peaks of the Nusselt number indicates the existence of an additional recirculating flow region, besides the primary one that is attached to the step corner, which has also been found by Armaly et al. (1983) (see enlarged sketch of Figure 1). Armaly et al., however, do not mention the presence of the above mentioned unstable flow configuration and report that the additional recirculating flow region disappears for $Re > 2300$. The latter discrepancy may be attributed to both the different ways of detecting the separation region in the two investigations and, to a lesser extent, to the different convergent nozzle being used.

At higher values of the Reynolds number, where only one peak of Nu is present, this peak slightly moves upstream with Re increasing up to about $x_{max}/s = 8$ for $Re = 7500$. Even if it is claimed that the turbulent range is characterized by an almost constant reattachment

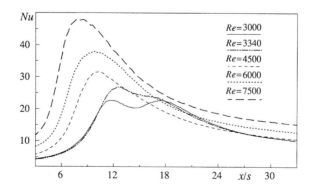

Figure 7   Nusselt number profiles on the centerline of the lower wall.

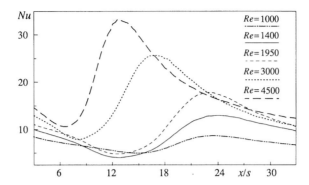

Figure 8 Nusselt number profiles on the centerline of the upper wall.

length, the measured value in this regime is in good agreement with the one obtained by Abbot and Kline (1962) in similar test-section geometry at higher Reynolds number and with Armaly et al. (1983) for a similar Reynolds number.

The presence of a recirculating-flow region on the wall opposite to the step (see enlarged sketch of Figure 1), caused by the strong adverse pressure gradient undergone by the flow, is evident from the diagram of Figure 8 where Nusselt number profiles on the centerline of the upper wall are reported. This region tends to move upstream and to shorten in its axial dimension for increasing Reynolds number. As already found by Armaly et al. (1983), this additional separation zone changes its location in accordance with the location changes of the reattachment point of the primary region of separation. Present heat transfer results show that this recirculating flow region seems to last up to a much higher Re with respect to Armaly et al. who report its disappearance for $Re > 6600$.

In Figures 9 through 16 the relief maps of the Nusselt number distributions on both sides of the channel downstream of the steps are reported. A substantial two-dimensionality of the flow is evident for $Re = 480$ even though some irregularities start to appear in the neighborhood of the reattachment zone on the upper wall. The bumpy aspect of the Nusselt number distribution (especially on the lower wall) for $Re = 1400S$, instead, bears evidence of the complex 3-D flow with several separated and reattached flow regions that occurs in the transitional regime. The map for $Re = 1400U$, which corresponds to the unstable flow configuration, seems to retain more of a certain degree of two-dimensionality. However, it has to be noticed that, as shown by Figures 12 and 14, no practically detectable difference is found for the convective heat transfer coefficient distribution in the presence of either one or two recirculating regions on the lower wall in the range $1400 < Re < 1950$.

At the highest Reynolds number ($Re = 7500$) an essentially 2-D flow is recovered; nevertheless some definite systematic transversal oscillations appear in the Nusselt number distribution on both sides of the channel. Since these oscillations show a very regular form that develops along the channel, they may be due to a pattern of vortexes which are superimposed to the main flow and produce the spanwise periodic variation of the convective heat transfer coefficient.

In the range $5000 < Re < 20000$ the flow configuration in the vicinity of the step is practically the one already related for $Re = 7500$. For further increasing Reynolds number, again a 3-D zone of separation is found especially in the proximity of the upper wall, across and just downstream of the step location. In particular, for $20000 < Re < 39000$ the thermograms seem to indicate there the presence of one or more vortexes, which seem to

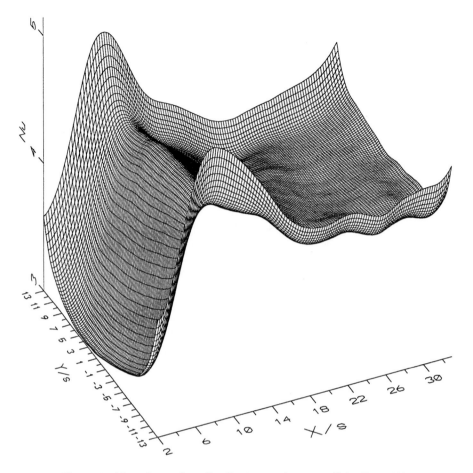

Figure 9   Nusselt number distribution on lower wall for Re = 480.

rotate about an axis normal to the two wide walls of the channel of a type similar to the one already described by Abbott and Kline (1962). It has to be stressed, however, that, while Abbott and Kline find always these vortexes on the step side and of small size, in this case there is strong evidence of much larger size vortexes only on the flat wall of the channel at the indicated Reynolds number range. The number and the position of such vortexes (see Figure 17b) appears generally random; the symmetry to the centerplane of the test section is lost, however, as shown by the thermogram of Figure 18b, only for about $Re = 35000$ do these vortexes seem to assume a regular symmetric pattern. In any case no evidence of such vortexes is found on the lower wall except for disturbances, most probably caused from them, which manifest after reattachment of the flow.

For further increasing Reynolds number, and as demonstrated by the thermograms of Figure 19 which are relative to $Re = 39500$, the regular pattern of vortexes indicated by Figure 18b tends rapidly to disappear and, around said Re value, any recirculating flow region on the wall opposite to the step vanishes. This behavior is shown by the monotonous trend of increasing temperatures in Figure 19b. However, it is still possible to notice a pronounced waviness, especially downstream of the reattachment zone, of the color bands in the thermograms of both the upper and lower walls. This waviness practically disappears for $Re > 42000$.

Finally, in the range $42000 < Re < 50400$, a 2-D turbulent flow is completely recovered; this behavior is evident in the thermograms of Figure 20 where temperature maps on both

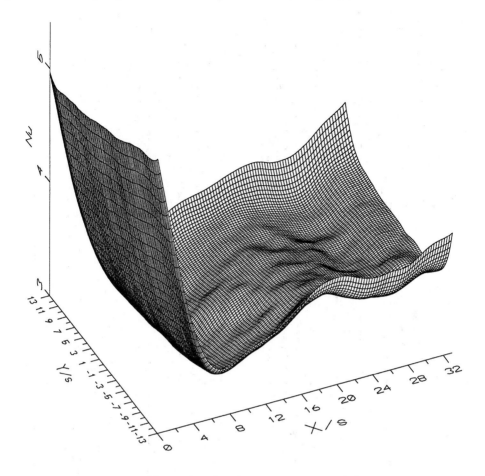

Figure 10   Nusselt number distribution on upper wall for Re = 480.

sides of the channel are shown for Re = 48000. As it may be seen by comparing, for example, the thermograms of Figures 17a and 20a, the temperature minimum on the lower wall is moved slightly downstream at the higher Re. In fact, for increasing Reynolds number from 35000 to 40000, the reattachment zone moves about two step heights downstream and the Nusselt number maximum decreases a little most probably because of the smaller curvature of the streamlines in the vicinity of the reattachment location. With further increase of the Reynolds number said maximum increases again and the reattachment zone remains fixed at the new location ($x_{max}/s \simeq 10$).

The relief maps of Figures 21 through 24 show the Nusselt number distributions on both sides of the channel for Re = 35000 and 48000 (i.e., corresponding to the thermograms of Figures 18 and 20) respectively. The map of Figure 22, which is relative to the regular pattern of vortices shown in the thermogram of Figure 18b, is quite impressive; it is hard to believe that such a high variation of the Nusselt number may occur spanwise on a flat wall even if in the vicinity of a step. Similar variations, although not showing a regular pattern such as the one in Figure 22, have been found in the range 20000 < Re < 42000. By comparing Figures 21 and 23 it is possible to see that the thermal reattachment zone has move slightly downstream for the highest Reynolds number and that, for this latter Re, the Nu peak looks wider. The map of Figure 24 shows a continuous decrease of the Nusselt number with an inflexion zone located slightly downstream of the thermal reattachment zone which is present on the opposite side of the channel.

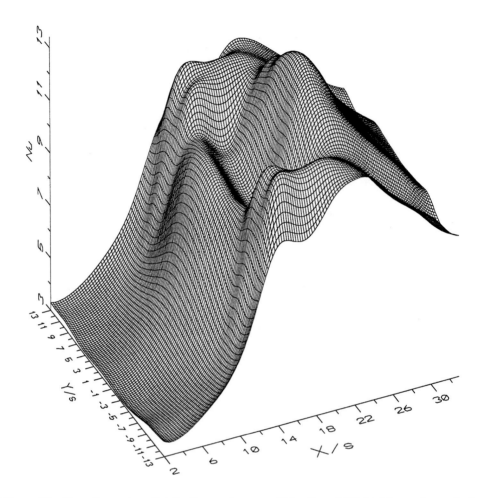

Figure 11   Nusselt number distribution on lower wall for Re = 1400 (stable configuration).

Conclusions

Surface flow visualization and heat transfer measurements in a backward-facing step flow are performed on both sides of the channel downstream of the step by making use of the *heated thin-foil* technique and by measuring temperature maps with an infrared scanning radiometer. The use of the radiometer turns out to be advantageous on account of its relatively good spatial resolution and thermal sensitivity.

In the laminar regime for $260 < Re < 500$ the flow is essentially 2-D and the reattachment length increases for increasing Reynolds number. In substantial accordance with Armaly et al. (1983), temperature maps show that, regardless of the high aspect ratio of the channel, a 3-D flow is present downstream of the step in the Reynolds number range $500 < Re < 5000$. The Nu distribution also shows an unstable flow configuration in the range $1400 < Re < 1950$ characterized by the presence, on the lower wall, of only one Nu peak as it occurs for $Re < 1400$.

In the range $1400 < Re < 3400$, the presence of two peaks in the Nu distribution on the lower wall centerline confirms the presence of an additional recirculating flow region previously reported by Armaly et al. (1983). At higher Reynolds numbers the only Nu peak

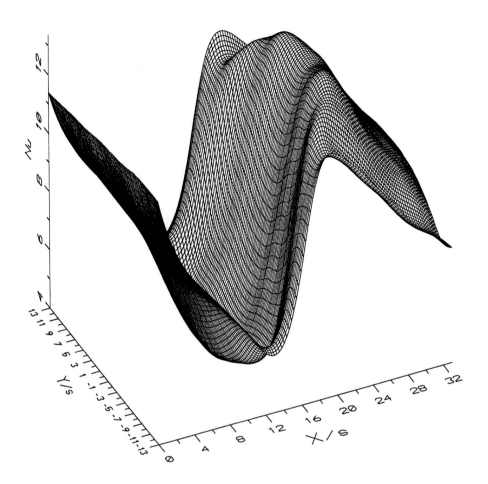

Figure 12   Nusselt number distribution on upper wall for Re = 1400 (stable configuration).

which is present moves slightly upstream until about $x_{max}/s = 8$; its location is in substantial agreement with the one obtained by Abbot and Kline (1962) and Armaly et al. (1983). The presence of a recirculating flow region on the wall opposite the step, also previously reported by Armaly et al., is confirmed by the heat transfer results as well. However, this region seems to last up to a much higher Reynolds number.

For 5000 < Re < 20000, the flow is again substantially 2-D with the presence of one separated region on each side of the channel; nevertheless some definite systematic transversal oscillations appear in the Nusselt number distribution on both sides of the channel.

In the range 20000 < Re < 39000, the presence of one or more vortexes, which seem to rotate about an axis normal to the two wide walls of the channel, is indicated in the proximity of the upper wall, across and just downsteam of the step. For increasing Reynolds number from 35000 to 40000 the reattachment zone moves about two step heights downstream and the Nusselt number maximum decreases slightly. For further increasing Reynolds number, said maximum increases again and the reattachment zone remains fixed at the new location ($x_{max}/s \simeq 10$).

At last, in the range 42000 < Re < 50400, a 2-D turbulent flow is completely recovered.

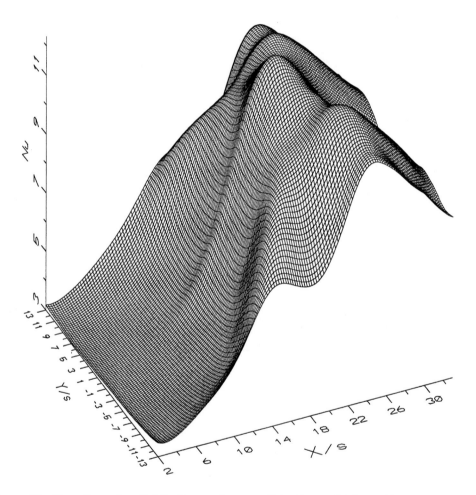

Figure 13 Nusselt number distribution on lower wall for Re = 1400 (unstable configuration).

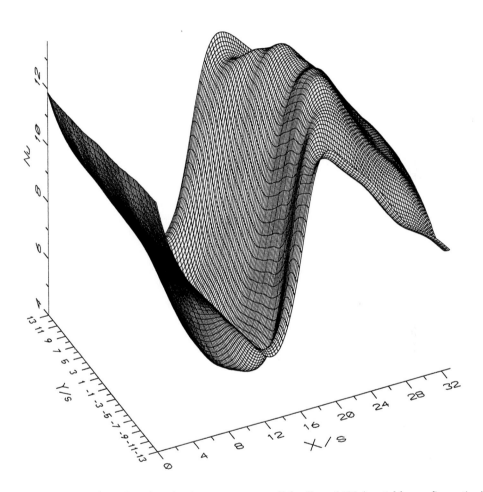

Figure 14   Nusselt number distribution on upper wall for Re = 1400 (unstable configuration).

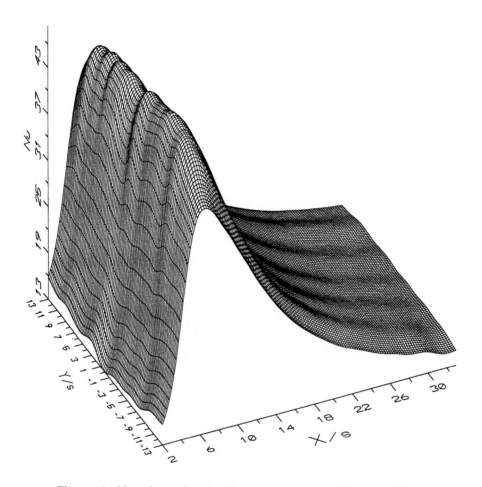

Figure 15   Nusselt number distribution on lower wall for Re = 7500.

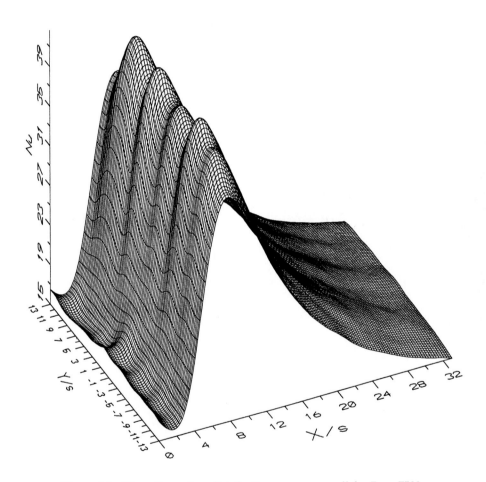

Figure 16   Nusselt number distribution on upper wall for Re = 7500.

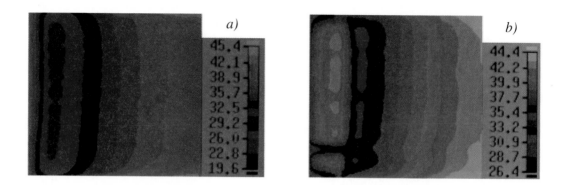

Figure 17   Temperature maps for Re = 20000: a) lower wall; b) upper wall.

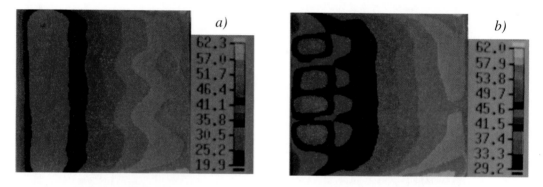

Figure 18   Temperature maps for Re = 35000: a) lower wall; b) upper wall.

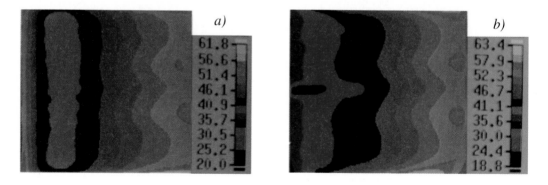

Figure 19   Temperature maps for Re = 39500: a) lower wall; b) upper wall.

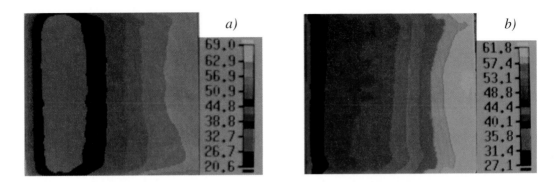

Figure 20   Temperature maps for Re = 48000: a) lower wall; b) upper wall.

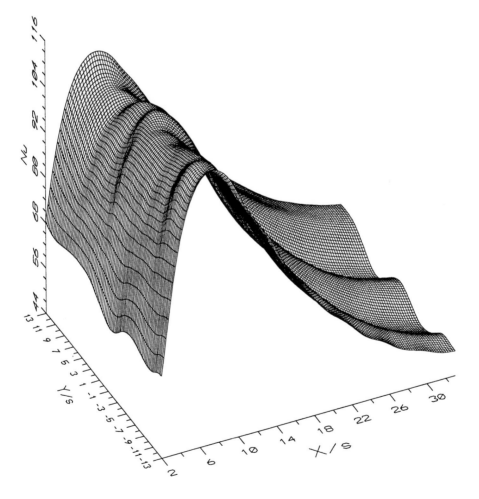

Figure 21   Nusselt number distribution on lower wall for Re = 35000.

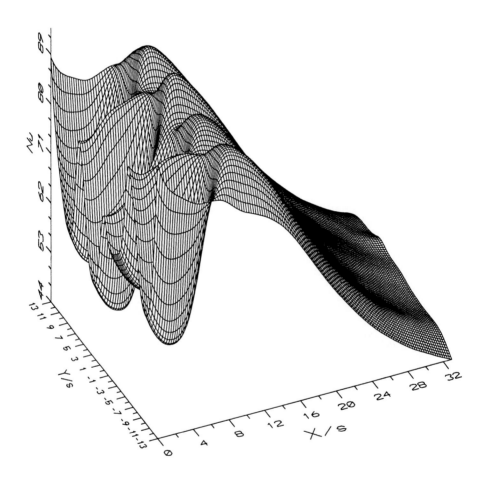

Figure 22   Nusselt number distribution on upper wall for Re = 35000.

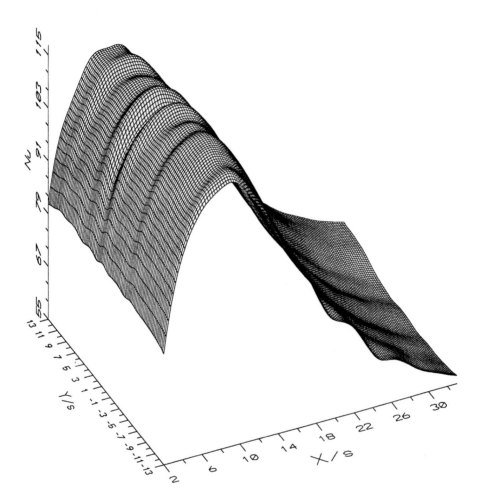

Figure 23   Nusselt number distribution on lower wall for Re = 48000.

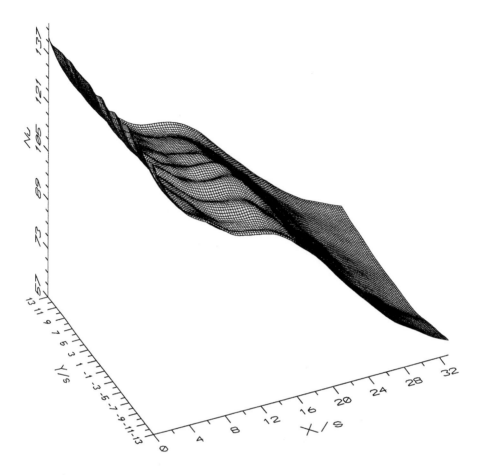

Figure 24   Nusselt number distribution on upper wall for Re = 48000.

References

Abbot, D. E. and Kline, S. J., 1962: Experimental investigation of subsonic flow over single and double backward facing steps, *Trans. ASME, J. Basic Eng.*, 84, pp. 317–340.

Armaly, B. F., Durst, F., Pereira, J. C. F., and Schönung, B., 1983: Experimental and theoretical investigation of backward-facing step flow, *J. Fluid Mech.*, 127, pp. 473–496.

Aung, W., 1983: An experimental study of laminar heat transfer downstream of backsteps, *J. Heat Transfer*, 105, pp. 823–829.

Aung, W. and Goldstein, R. J., 1970: Temperature distribution and heat transfer in a transitional separated shear layer, in *Heat Transfer*, Grigull, U., Hahne, F., Eds., Vol. 2, pp. 1–11, Elsevier, New York.

Cardone, G., di Leva, O. M., and Carlomagno, G. M., 1993: Heat transfer in a backward-facing step flow, *Proc. II Int. Conf. Fluid Mech.*, pp. 970–975, Peking University Press, Beijing.

Carlomagno, G. M. and de Luca, L., 1989: Infrared thermography in heat transfer, in *Handbook of Flow Visualization*, Yang, W. J., Ed., Ch. 32, pp. 531–553, Hemisphere, New York.

de Luca, L., Carlomagno, G. M., and Buresti, G., 1990: Boundary layer diagnostics by means of an infrared scanning radiometer, *Exp. in Fluids*, 9, pp. 121–128.

Kottke, V., 1983: Strömung, Stoff-, Wärme- und Impulsübertragung in lokalen Ablösegebieten, *Fortschr. Ber. VDIZ*, Reihe 7, Nr. 77.

Scherer, V., Wittig, S., Bittlinger, G., and Pfeiffer, A., 1993: Thermographic heat transfer measurements in separated flows, *Exp. in Fluids,* 14, pp. 17–24.

Seban, R. A., 1964: Heat transfer to the turbulent separated flow of air downstream of a step in the surface of a plate, *Trans. ASME, J. Heat Transfer,* 86, pp. 259–270.

Sparrow, E. M., Kang, S. S., and Chuck, W., 1987: Relation between the points of flow reattachment and maximum heat transfer for region of flow separation, *Int. J. Heat Mass Transfer,* 30, pp. 1237–1245.

Vogel, J. C. and Eaton, J. K., 1985: Combined heat transfer and fluid dynamic measurements downstream of a backward-facing step, *J. Heat Transfer,* 107, pp. 923–929.

chapter eight

A Technique for Visualization of the Turbulent Vortex Ring

V. F. Kopiev, M. Yu. Zaitsev, L. P. Guriashkin, and V. A. Yakovlev

Abstract *— The visualization possibility of nonstationary phenomena in the turbulent vortex ring core was investigated using an equipment set usually applied in optics-physics investigations for high-speed cinematography and flash photography. Vortex ring section photographs taken according to the impulse light-screen technique, as well as volume images obtained by means of a shearing interferometer, are presented. Cinema-grammes of the vortex ring core evolution performed with the image frequency of about 10^4 frames/s are given.*

Discussion

The subject of the present study is a gas turbulent flow well known in the field of hydrodynamics as a vortex ring (Lavrent'ev and Sabat 1980, Maxworthy 1974). A vortex ring structure allows one to distinguish between three characteristic flow fields: a toroidal vortex core, an ellipsoidal field surrounding the core and moving together with it (the so called ring "atmosphere"), and the outer flow. At high Reynolds number (Re > 1000) the flow appears turbulent throughout the whole ring "atmosphere", with the exception of the slowly altering vortex core in which the turbulence is suppressed (Vladimirov et al. 1980).

It was experimentally found (Zaitsev et al. 1990) that vortex ring movements are followed by sound radiation that can be associated with its core vibrations (Kopiev and Leont'ev 1987).

The present work studied possibilities of visualization of fast, nonstationary processes in the vortex ring core and methods of high-speed and flash photography were worked out. An equipment set used in optics and physics investigations for high-speed cinematography and flash photography was applied here for realization of the questions raised. They are the following:

- specially developed flash white-light sources — one with impulse duration 10^{-5} s and another with extended duration up to 3×10^{-3} s to work with high-speed cameras; pulsed ruby laser (impulse duration 30 ns, impulse energy 0.5 Joule)
- industrial high-speed drum camera with frame rate up to 20000 frames/s
- a mirror shearing interferometer

The equipment indicated above allowed us to develop a technique which permitted us to carry out high-speed photographic recording of vortex core processes with the image frequency up to 10^4 frames/s and to make single photographs with the exposure time up to 10^{-8} sec.

For vortex ring production, the piston vortex generator with a removable nozzle set was used. The characteristic Reynolds number predicted with reference to the initial vortex ring velocity $U = 25$ m/s and the generator nozzle diameter $d = 5 \times 10^{-2}$ m formed $Re = Ud/v = 8.3 \times 10^4$ which corresponds to turbulent flow conditions.

Flow field structure features associated with the existence of the distinct boundary between the laminar core and the turbulent "atmosphere", and with fluid particle transition from the "atmosphere" into the wake, require the development of adequate visualization methods.

Two visualization types were used: (a) smoke visualization in reflected light; (b) visualization with helium or ethyl steam admixture by means of the shearing interferometer.

In the first case heavy smoke particles are pushed out of the core and the contrast is determined by the difference of smoke concentrations in the uncolored core and the vortex ring "atmosphere". This method allows for visualization of the vortex core at a distance not more than 40 diameters from the nozzle exit plane of the vortex generator. At a greater distance the smoke concentration in the core and in the "atmosphere" is equalized due to smoke transition from the "atmosphere" into the wake.

Visualization with the help of the interferometer when helium and ethyl steam are used is effective at a distance more than 20–30 diameters from the nozzle exit plane because the core is masked by the helium-colored small scale turbulence of the "atmosphere" in the initial part of the vortex trajectory. In this case visualizing substance transition from the "atmosphere" into a wake is a positive factor as it reveals the helium-colored core.

It is possible to visualize pure rings without any admixtures at a distance of up to 30 diameters if one uses the naturally originating density reduction in the core. Only the central line of the vortex core appears visible where the density is minimal. The flow intensity weakens as it moves away from the nozzle exit plane and the density ratio appears insufficient to receive an image even with the help of an interferometer.

Photography and cinematography of smoke rings followed the scheme shown in Figure 1. The ring was illuminated by an impulse light source at the moment when it

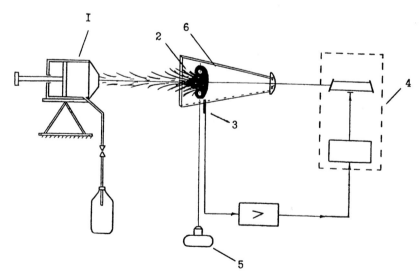

Figure 1 Experimental installation scheme for photo/cinema registration of a smoke ring lateral section using a light-screen method: (1) vortex generator; (2) vortex ring; (3) trigger microphone; (4) flash white-light source; (5) photo/cinema camera; (6) light sheet.

passed a fixed point in the space. Synchronization was ensured by a microphone which produced a trigger electric impulse at the moment the vortex core passed over its surface. Registration was performed in a dark room with open camera shutters. The microphone signal started a flash light source action and photos or cinema-grammes of rings were taken; after those actions the shutters were closed.

Single photos were taken according to a light-screen technique (a light "knife", a laser "knife"). The planar white-light sheet was produced by a flash lamp with a cylindrical lens in front of it. The plane was 5×10^{-3} m thick. A smoke ring entering the registration field was cut by a light "knife" in longitudinal or lateral directions which allowed one to obtain characteristic ring sections — a side view (Figure 2a) and a front view (Figure 2b).

In the case of a laser "knife", a luminous sheet was produced by pulsed laser (impulse duration $\simeq$ 30 ns, λ = 693.4 nm). Short-time exposure, slight thickness of the luminous plane (10^{-3} m), and highly monochromatic radiation provide a clearer structure of the vortex ring core. Longitudinal and lateral sections are presented in Figures 2c and d. Moreover, the large radiation energy in the impulse allows one to take photos at a distance of 40 or more diameters from the nozzle exit plane, when the smoke particle concentration in the ring "atmosphere" becomes insignificant.

Cinema-grammes of the vortex ring core evolution are obtained with the help of a high-speed drum camera which obtains up to 20000 frames/s (the frame size: 7.5×10 mm; the number of frames: up to 256; focusing limit: from 1m; the film width: 35 mm; the picture distribution: in two rows with a shear).

Cameras of this class have high loss of light in an optical path. Moreover, the photography was performed by collecting the reflected light with a low reflection coefficient of the object. This is why a specially developed high-intensity flash source of the extended duration $\tau = 3 \times 10^{-3}$ s was used for the registration field illumination. The photography was performed with two different viewing angles according to a scheme similar to the individual photography.

In this case the process registration time was determined by the flash duration. At the image frequency 10000 frames/s there are 32 pictures on the film (16 pictures in each row). The contrast image is obtained only when a highly-sensitive film is used and when the distance from the nozzle exit plane is up to 30 diameters (Figures 3a and b). The frame rate increases up to 15000 frames/s and transition to larger distances from the nozzle leads to a significant reduction of the negative's density.

For a more comprehensive flow picture in the neighborhood of the laminar core of a vortex ring, the visualization technique based on a shearing interferometer was used. This technique is successfully used in many aerodynamic experiments for gas flow visualization (Holder and North 1963) and it is especially convenient for two-dimensional (2-D) flow visualization. In the case of a vortex ring the image is a projection of one of the density gradient components over the whole ring volume onto the exposure plane and this impedes the result interpretation.

In the present investigation a visualization system with a mirror shearing interferometer is used in the most simple (auto-collimation) version. The system consists of three fundamental elements: A four-mirror interference-shadow device with a lateral shear of the wave fronts, the main spherical mirror, and a control panel. The vortex ring visualization scheme on the base of the interferometer and cameras is presented in Figure 4.

With the help of the visualization system described above photos (Figures 2d and e) and high speed cinema-grammes of a vortex ring were taken (Figures 5, 6).

Thus, the photos taken during the work give an idea of the momentary states of the vortex ring core and the developed registration technique with the maximum achieved frame rate allowed for investigations of fast dynamic processes in the core which do not exceed the frequency 1.5 to 2 kHz.

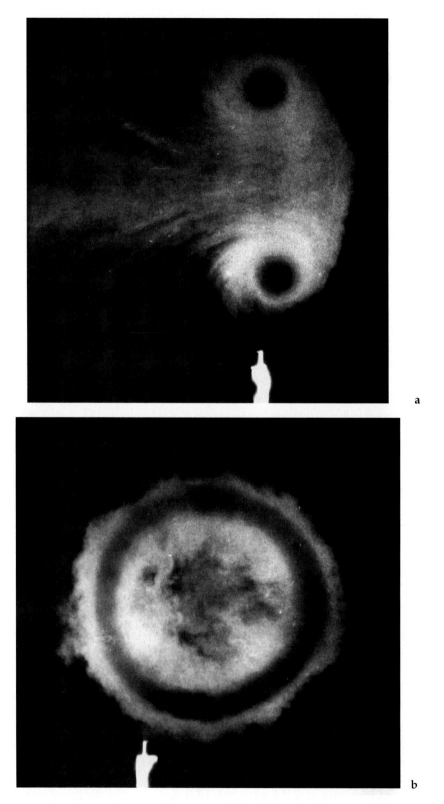

Figure 2   Vortex ring photographs, Re = 8.3 × 10⁴: a, b, c, d — vortex ring, longitudinal and lateral sections; e — vortex ring with helium admixture, a front view; f — pure vortex ring, a 3/4 front view. (There is a trigger microphone at the bottom of the picture.)

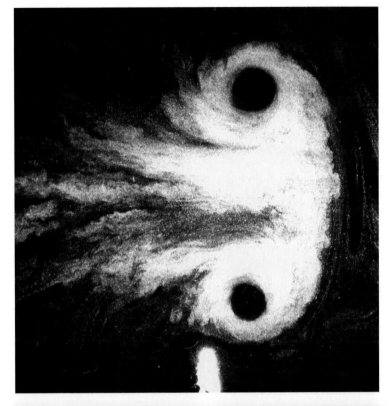

c

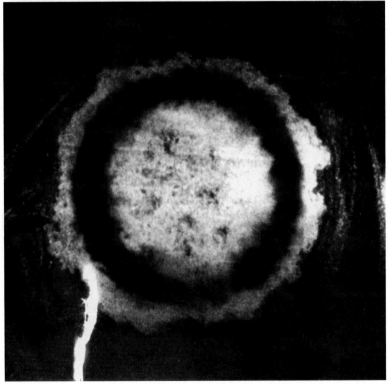

d

Figure 2 (continued)

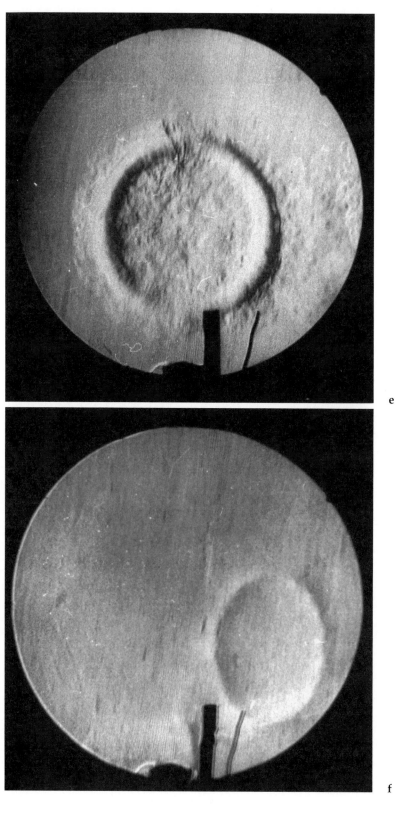

e

f

Figure 2 (continued)

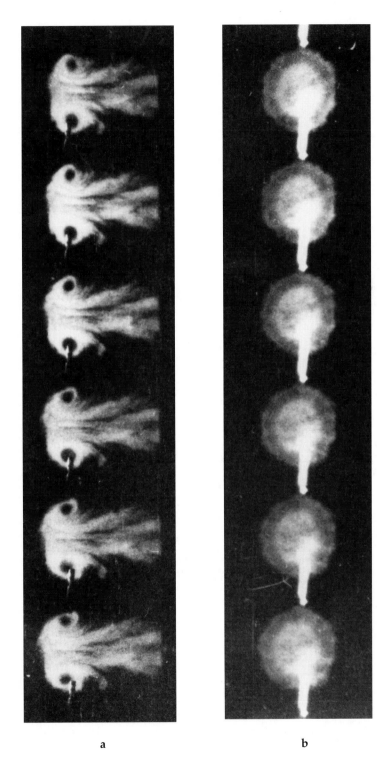

a   b

Figure 3   Cinema-grammes of the vortex ring core evolution, $\eta = 10^4$ frames/s: a — a side view; b — a front view.

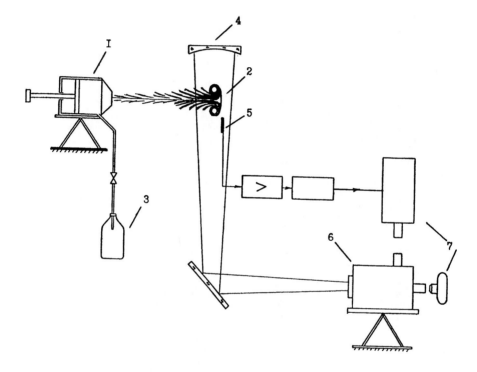

Figure 4   The visualization scheme of a vortex ring with the help of a shearing interferometer: (1) vortex generator; (2) vortex ring; (3) tank with helium; (4) spherical mirror; (5) trigger microphone; (6) shearing interferometer; (7) photo/cinema camera.

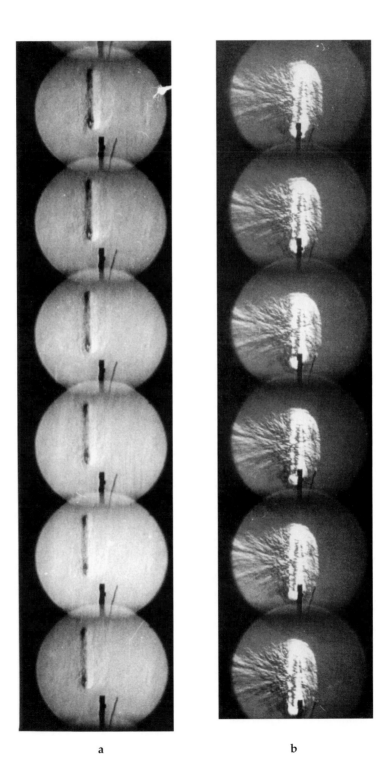

a   b

Figure 5 Cinema-grammes of vortex ring movements, $\eta = 10^4$ frames/s, a side view: a — visualization by ethyl steams; b — visualization by helium admixtures.

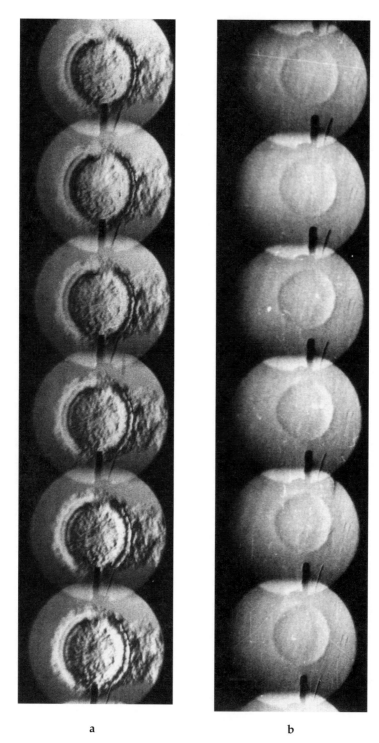

a   b

Figure 6 Cinema-grammes of vortex ring movements, $\eta = 10^4$ frames/s, a front view: a — with helium admixtures; b — pure.

The main results of this work were presented at the conference "Optical methods of analyzing flows" (Kopiev et al. 1991).

References

Holder, D. W. and North, R. J.: Schlieren methods, *NPL Notes Appl. Sci.*, No. 31, (1963).

Kopiev, V. F. and Leont'ev, E. A.: Radiation and scattering of sound from a vortex ring, *Izv. Akad. Nauk SSSR Mech. Zidk, Gaza*, Vol. 22, No. 3, pp. 83–95, (1987).

Kopiev, V. F., Zaitsev, M. Yu., Rybakov, V. I., Bezmenova, T. N., Guriashkin, L. P., Komissarova, A. F., Frolova, A. D., Yakovlev, V. A.: A technique and techniques for visualization and high-speed photography of non-stationary processes in the core of a turbulent vortex ring, *Izv. Sib. Otd. Akad. Nauk SSSR, Ser. Techn. Nauk*, No. 6, pp. 54–61, (1991).

Lavrent'ev, M. A. and Sabat, B. V.: Effets hydrodynamiques et modeles mathematiquvse, *Progress. Moskva.*, (1980).

Maxworthy, T.: Turbulent vortex rings, *J. Fluid Mech.*, Vol. 64, pp. 227–239, (1974).

Vladimirov, V. A., Lugovtsov, B. A., Tarasov, V. F.: Turbulence suppression in the cores of concentrated vortices, *Appl. Mech.*, No. 5, pp. 69–76, (1980).

Zaitsev, M. Yu., Kopiev, V. F., Munin, A. G., Potokin, A. A.: Sound emission by the turbulent vortex ring, *Dokl. Akad. Nauk SSSR*, Vol. 312, No. 5, pp. 1080–1083, (1990).

chapter nine

Condensation Phenomena in Laval Nozzles*

Günter H. Schnerr and Jürgen Zierep

Institut für Strömungslehre und Strömungsmaschinen
University of Karlsruhe, Germany

Abstract *— The microscopic and macroscopic effects in nozzle flows with nonequilibrium phase transition have been investigated in theory and experiment. A colored schlieren optical system has been used for visualization of the complex interaction of the flow and the condensation process. The numerical algorithm solves the system of the Euler equations coupled to the homogeneous nucleation rate and droplet growth equations using a diabatic finite volume method. The reservoir vapor pressure, the cooling rate of the expansion, and the nozzle wall curvature that appear to be the dominating parameters are varied to demonstrate all relevant two-dimensional (2-D) phenomena in flows with sub- and supercritical heat addition. In 2-D flows the nucleation process shows the most sensitive interaction with the gradients in the flow field. The resulting complex structures explain the macroscopic gasdynamic effects on pressure and density. Additional nucleation zones in supersonic expansions and subsonic heating fronts for higher nozzle wall curvatures are discussed in detail. Quantitative comparison of numerical and experimental flow visualization allows the emphirical determination of unknown quantities of the classical nucleation theory.*

Introduction

Phase transition normally takes place in thermodynamic equilibrium. In flows this holds true as long as the characteristic time scale is small enough to maintain the vapor and liquid phase in equilibrium. Obvious examples are atmospheric cloud formation and falling rain. In turbomachinery and transonic wind tunnels, e.g., cryogenic test facilities, the speed of flow may increase up to the speed of sound and higher, simultaneously the fluid cools down with gradients of about 0.1 to 1°C/μs (Hall 1979, Kilgore 1989, Schnerr and Dohrmann 1990a, 1991, Wegener 1987, 1991). Under these conditions the condensing vapor component may become highly supersaturated, i.e., the pressure and temperature lie in a range where normally the vapor would be liquified or solidified. If no surfaces are available from particles or aerosols for the deposition of the higher liquid phase, the condensation nuclei must be formed in the pure vapor phase by kinetics of phase change. Volmer and Weber

*This work is dedicated to Peter P. Wegener, Harold Hodgkinson Professor Emeritus of Yale University, on the occasion of his outstanding contribution to this subject.

(1926) gave the first expression for the rate of formation of nuclei of the critical size. Maximum values in flows presented in this paper for the homogeneous nucleation rate are of the order of 10^{25} [m^{-3} s^{-1}] and much higher as particle concentrations of heterogeneous systems. The smallest nuclei at the so called Wilson point consists of only a few molecules, typically 10 to 20, having a characteristic diameter of less than one order larger than the molecule size of about $3 \cdot 10^{-10}$ m.

For atmospheric reservoir conditions the kinetics requires a mean supersaturation of the water vapor of at least about 50 K, but only 40 K to accelerate the flow to Mach number unity. This reveals the important fact that the macroscopic latent heat release to the flow falls into the supersonic region, i.e., the divergent supersonic part of a Laval nozzle (Oswatitsch 1941) or in transonic flows around airfoils inside of the local supersonic area. For the same reason the so-called condensation onset Mach number is always around unity and in the range between 1.1 to 1.4. Supersonic heat addition compresses the flow toward to the critical state. With respect to the maximum mass flow rate, just at Mach number one and the equivalence of mass and energy supply in compressible flows, we immediately understand the sensitive reaction of transonic flows to the smallest rates of heat addition and the occurrence of phenomena like "condensation shocks" and periodic shock oscillations. However, the meaning of the expression "condensation shock" is not very clear and only understandable as a rough interpretation of the first visualization of this phenomena by Prandtl (1935) and Hermann (1942). A more accurate classification of stationary condensation phenomena subdivides in flows with sub- and supercritical heat addition (Zierep 1971, 1974, 1980, 1990). An asymptotic predictive method for one-dimensional (1-D) nozzle flows with nonequilibrium condensation that reveals the detailed structure of the condensation zones is also available from the work of Clarke and Delale (1986 a, b) and Delale (1990). Complex boundary conditions, i.e., the curvature of individual streamlines cause additional structures like sub- and supersonic heating fronts (Schnerr 1989).

Stationary and nonstationary nonequilibrium wet steam flows have been investigated by Moheban and Young (1984), a fundamental study of stationary and moving shock waves in wet steam can be found by Guha and Young (1990) and Young and Guha (1991). Zierep and Lin (1967, 1968) published the first similarity laws for the stationary onset of homogeneous condensation and for the onset of periodic oscillations in Laval nozzles.

In the following, all relevant phenomena in stationary nozzle flows are collected and visualized at the macroscopic and microscopic scale. Original color schlieren pictures demonstrate the macroscopic gasdynamic effects of the latent heat release on the density. Numerical computations allow the analysis of the phase transition process from the beginning at the microscopic scale. The most important parameters of the problem, the reservoir conditions and the cooling rate of the flow, are varied and 2-D effects are demonstrated for different nozzle wall curvatures but constant cooling rates. The fluid is humid air, i.e., a mixture of one condensable vapor and a mixture of inert carrier gases.

Method of Measurements

The experiments are performed in the atmospheric supersonic indraft wind tunnel of our institute. Without internal fittings the maximum cross-sectional area of the test section is 230×50 mm² (height × width). The nozzles used had a total height of 120 mm and 60 mm at the critical cross-section. The area of visualization is limited by the windows mounted at the sidewalls of the wind tunnel that are 220 mm in diameter. The flows were visualized by means of a schlieren optical system with an exposure time of 1 μs by a single spark of a Strobokin spark light source (Impulsphysik Hamburg). A prism without deflection of the axis of refraction (opening angle of 2°47′) was located at the focal point of the first mirror, and — instead of the knife edge — a slit was mounted at the second focus location. The position and width of the slit, in this case 3 mm, determine the undisturbed spectral color of the background, in Figure 1 from red-orange to turquoise. The schlieren slit was always adjusted normal to the nozzle axis; this visualizes the density gradients in main flow direction. The film material was a conventional daylight slide quality with a sensitivity of 200 ASA.

Effects of Nozzle Wall Curvature and Vapor Pressure in the Reservoir

Figure 1 is a collection of 9 original schlieren pictures of Laval nozzle flows with homogeneous condensation in the right hand supersonic part, flow direction is from the left. Each column of this figure represents a constant nozzle geometry. In the vertical direction — as the main effect — the reservoir vapor pressure increases from top by a factor of two in the corresponding nozzle. Along the rows the reservoir conditions are approximately kept constant, but the nozzle geometry varies.

The circular arc nozzle No. 1 in the first column has a total throat height of $2y^* = 60$ mm and a constant radius of wall curvature of $R^* = 200$ mm. Nozzle No. 3 in the third column has the total throat of $2y^* = 120$ mm and a constant radius of wall curvature of only 100 mm, i.e., the term $\sqrt{y^* R^*}$ which is the characteristic length of the expansion and with that the predominating cooling rate $(-dT/dx)^*$, respectively $(-dT/dt)^*$ are kept constant (Wegener 1964).

$$\left(-\frac{dT}{dx}\right)^* = 2\frac{\gamma-1}{(\gamma+1)^{3/2}}\frac{1}{\sqrt{y^* R^*}}\,T_{01}$$

$$\left(-\frac{dT}{dt}\right)^* = 2^{3/2}\frac{\gamma-1}{(\gamma+1)^2}\sqrt{\frac{\gamma R}{y^* R^*}}\,T_{01}^{3/2} \tag{1}$$

Here T_{01} is the temperature at the reservoir, t the time, γ the specific heat ratio, $\mathbf{R}$ the individual gas constant, and x,y are the Cartesian coordinates. As a good approximation these derivatives may be taken at the critical throat position at Mach number unity, indicated by the superscript*. Nozzle No. 2 in between has the same total throat height as nozzle No. 3 ($2y^* = 120$ mm) but the larger radius of wall curvature at the throat of the nozzle No. 1 ($R^* = 200$ mm). This reduces the cooling rate about 30%, see Table 1.

Table 1   Theoretically and Experimentally Determined Critical Temperature Gradients of the Tested Circular Arc and Hyperbolic Nozzles

| | | | | | \multicolumn{3}{c}{$(-dT/dx)^*_{axis}$[°C/cm] $T_{01} = 293$ K} | | |
| | | | | | | 2-D | |
Nozzle	y^* (mm)	R^* (mm)	y^*R^* (mm²)	y^*/R^*	1-D	Theory	Experiment
1	30	200	6000	0.1500	8.14	7.64	7.33
2	60	200	12000	0.3000	5.76	5.18	5.13
3	60	100	6000	0.6000	8.14	7.20	6.85

Comparing columns 1 and 3 demonstrates the development of 2-D effects caused by the nozzle wall curvature at constant cooling rates. Columns 2 and 3 depict the effects of the nozzle wall curvature as before, but now for different cooling rates and for constant nozzle throat heights. From the comparison of columns 1 and 2 we recognize the effect of different throat heights for constant nozzle wall curvatures, i.e., different cooling rates as well.

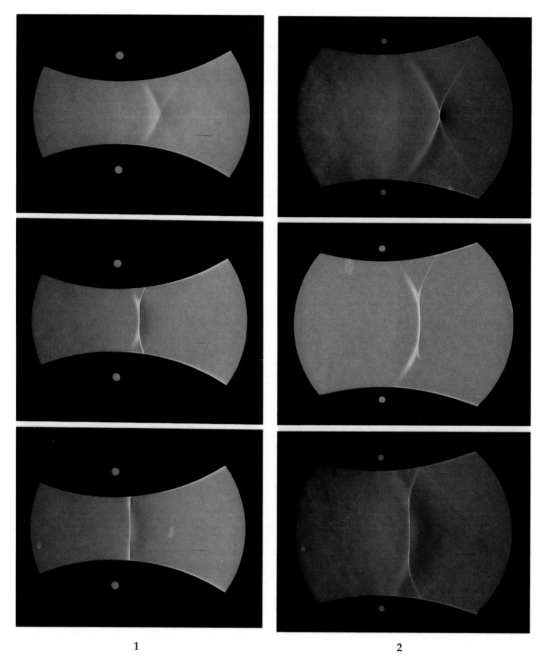

1   2

Figure 1   Laval nozzle flows with homogeneous condensation of water vapor in moist air, flow is from the left. The two holes indicate the throat position. First column: nozzle No. 1; second column: nozzle No. 2; third column: nozzle No. 3 (see Table 1). Atmospheric supply: reservoir pressure p_{01} = 1bar, reservoir temperature 290.4K ≤ T_{01} ≤ 303.9K. First row: mean value of the relative humidity in the reservoir ϕ_0 = 35.8%, "x-shock" and dominating subcritical heat addition in supersonic flow. Second row: mean value of the relative humidity in the reservoir ϕ_0 = 56.8%, mixed sub- and supercritical heat addition; subsonic heating fronts on the right. Third row: Mean value of the relative humidity in the reservoir ϕ_0 = 68%, pure supercritical and mixed sub- and supercritical heat addition.

Effect of the Vapor Pressure in the Reservoir

First row of Figure 1: — Low vapor pressures or low values of the relative humidity below 40% in the upper row create the classical so-called "x-shock". Under these conditions the

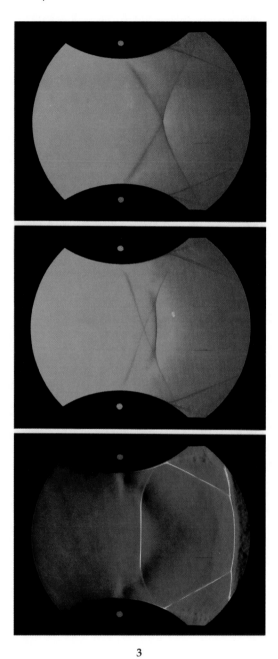

3

macroscopic onset of heat addition approximately follows an isoline of the density or temperature. The bright sickle-shaped area behind indicates the supersonic compression (nozzle No. 1). The compression is weak and the flow still remains supersonic. Convergent characteristics emerging from the compression zone intersect and form the weak oblique shocks. This is the characteristic behavior of subcritical heat addition. However, nozzle No. 2 shows just the beginning of a small normal shock in the center. Constant vapor pressure in the supply and the lower cooling rate of this nozzle decrease the supersaturation at the collapse of the metastable state at the Wilson point and the heat addition starts at a lower onset Mach number. Locally the flow becomes just supercritical, i.e., the stationary flow requires a normal shock inside of the zone of heat addition.

Second row of Figure 1: — The formation of local supercritical flow becomes more evident in the second row. Here the relative humidity lies between 50 and 60%. Each of the more complicated shock structures consists of an extended normal shock and oblique shocks. The results from nozzles 1 and 2 are similar. Along the streamlines near the axis of symmetry there is no significant curvature, and if at all, in this region the heat addition becomes supercritical for the first time. This means that already a small percentage of the condensable vapor makes the flow to be thermally choked just behind the onset of the heat release. To overcome this singular situation a normal shock is built up which first reduces the speed of flow to subsonic, then it accelerates again from sub- to supersonic continuously without any singularity, driven by the latent heat release of the condensation process. In principle this normal shock caused by supercritical heat addition tends to diminish the condensation process. The temperature jump across the shock reduces the supersaturation and with that the nucleation and droplet growth rate. Although this shock originally is caused by heat addition from homogeneous condensation, its characterization as a "condensation shock" is not correct and just the opposite of that is what happens in the flow.

In constant area flows (ducts) stationary supercritical heat addition is not possible. Choking under these conditions means that the Mach number becomes unity just after the heating zone. Simultaneously, this is the limit of stationary flows. Higher rates immediately create nonstationary flows (Bartlmä 1966, 1975).

Third row of Figure 1: — The dominating effect shown in the third row is the variation of the relative nozzle wall curvature y^*/R^* which increases from the left from 0.15, 0.3 to 0.6. Consequently nozzle No. 3 on the right hand side depicts the most intense 2D-effects. The vapor pressure respectively the relative humidity in the reservoir is about twice the amount of that of the first row. The nearly normal shock in nozzle No. 1 at the left hand side seems to indicate a complete 1-D flow situation which is not the correct conclusion. From other pictures of the same nozzle it follows that the flow variables vary in a 2-D manner. Thus, the stationary straight shock is a singular configuration in 2-D flows with supercritical heat addition as a whole. At first the speed of flow accelerates at the throat to supersonic, then it reduces to subsonic across the shock, and accelerates once again to supersonic. Behind the shock there is a local subsonic area bounded by a second sonic line extending from the upper to the lower nozzle wall.

Near the wall the structure may still remain subcritical by the formation of oblique shocks. New phenomena occur from the development of subsonic heating fronts. This will be discussed in more detail in the subsequent section. The additional shocks downstream in nozzle No. 3 are caused by the higher back pressure at the exit of the nozzle. They show the intensive interaction of strong shocks with the side wall boundary layer in the test section downstream and upstream of the shock. Obviously, these interesting effects are completely separated from the condensation zone.

2-D Effects of the Nozzle Wall Curvature

Third column of Figure 1: — The flow near the plane of symmetry shown in the middle of this column is disturbed by compression waves extending from the onset of heat release near the well-curved nozzle walls. The point of intersection of this "x-shock" is now located upstream of the condensation onset in this region. Static pressure measurements already indicate a deviation from the adiabatic flow and may be misinterpreted as the condensation onset. There the onset fronts obviously are subsonic heating fronts, i.e., the latent heat is added in the supersonic flow where the velocity component normal to the heating front is still subsonic. The existence of this wave propagation in 2-D steady flows was previously pointed out by Bratos and Meier (1976). Under these circumstances, measurements of pressure or density by interferometry cannot be used for the experimental determination of the onset of condensation phenomena. Numerical analysis of the flow inside of local supersonic areas over airfoils reveals the important fact that this behavior dominates the entire nucleation and condensation processes in these flows. Finally, per definition, heating fronts from condensation processes in Prandtl-Meyer expansions locally are to be expected as sonic heating fronts.

In conclusion, in numerical calculations of the more complex 2-D flows we define for the accurate determination of the macroscopic condensation onset a certain amount of the condensate mass fraction, about 1% of the maximum value. In experiments, only light scattering methods may give more correct information.

Numerical Simulation and Visualization

Numerical modeling is based on the Euler equations linked with the classical nucleation theory of Volmer (1939) and the microscopic droplet growth law of Hertz-Knudsen. An improved time-dependent diabatic finite volume method is developed and applied to calculate stationary flows. The equations for mass, momentum, and energy are used in the conservation form:

$$\frac{\partial U}{\partial t} + \frac{\partial F}{\partial x} + \frac{\partial G}{\partial y} = S \tag{2}$$

In Equations (2) and (3) the source term S represents the latent heat release of the phase change process. Here U is the vector of the dependent variables, F and G are the conservative flux vectors:

$$U = \begin{pmatrix} \rho \\ \rho u \\ \rho v \\ \rho e \end{pmatrix} \quad F = \begin{pmatrix} \rho u \\ \rho u^2 + p \\ \rho u v \\ (\rho e + p)u \end{pmatrix} \quad G = \begin{pmatrix} \rho v \\ \rho u v \\ \rho v^2 + p \\ (\rho e + p)v \end{pmatrix} \quad S = \begin{pmatrix} 0 \\ 0 \\ 0 \\ \rho \dfrac{dQ}{dt} \end{pmatrix} \tag{3}$$

ρ is the density, p the static pressure, e the total energy per unit mass, Q the latent heat release, and u and v are the components of the velocity vector. The main issue of any nucleation theory is the determination of the Gibb's free enthalpy ΔG^* for the formation of the critical nucleus. Here, the asterisk indicates the metastable size of a cluster which usually is called a nucleus. Fortunately, from the time scale of the transonic expansion flow and the internal characteristic time scale for the formation of the nucleus, it follows that the stationary nucleation theory is still appropriate for the modeling of our nonequilibrium condensation process. The nucleation rate J per unit volume and unit time and the droplet growth rate may be written as:

$$J = \sqrt{\frac{2}{\pi}}\, \sigma_\infty\, m^{-3/2} \frac{\rho_v^2}{\rho_c} \exp\left(-\frac{\Delta G^*}{kT}\right)$$

$$\Delta G^* = \frac{16}{3}\pi \left(\frac{m}{\rho_c \ln(s)kT}\right)^2 \sigma_\infty^3 \tag{4}$$

$$\frac{dr}{dt} = \frac{\alpha}{\rho_c} \frac{p_v - p_{s,r}}{\sqrt{2\pi R_v T}} \tag{5}$$

In Equations (4) and (5) σ_∞ is the surface tension of a plane surface, m the mass per molecule water, ρ_v and ρ_c the density of the vapor and the condensate, k the Boltzmann constant, s the supersaturation ratio, $p_{s,r}$ the saturation vapor pressure of a droplet of the radius r, α the mass accomodation coefficient, and $\mathbf{R}_v$ the individual gas constant of the vapor. The surface tension σ_∞ in Equation (4) and the mass accomodation coefficient α in Equation (5) are unknown in the actual range for temperatures of about 220K. The comparison of extended nozzle experiments and numerical calculations yields empirical relationships for the surface tension and an approximately constant mass accommodation coefficient α equal to unity. The 2-D, diabatic finite volume method is explicit in calculating both the dependent variables and the condensate mass fraction. Nonreflecting boundary conditions complete the flow variables at the elliptic inflow boundary. More details of the numerical method are already given by Dohrmann (1989) and Schnerr and Dohrmann (1989, 1990b).

Viscosity effects are totally insignificant in these flows. Over all, the flow is accelerated without recognizable local or total separation within the diabatic compression zone. Numerical calculations of the homogeneous condensation process inside turbulent boundary layers confirmed this experimental observation (Schnerr et al. 1992). What remains is only a quantitatively small correction due to the displacement effect of the boundary layer.

Three experiments from Figure 1 were recomputed and analyzed for the nucleation rate J, the condensate mass fraction g/g_{max}, and the static pressure disturbance $\Delta p = p_{diabatic} - p_{adiabatic}$ normalized by the reservoir pressure p_{01} of the mixture, always about 1bar in these examples. Figure 2 shows the color-chart of these quantities. Dark blue represents negligible low values. Relevant values of the nucleation rate with respect to macroscopic gasdynamic phenomena concentrate in the orange- and red-colored domain, i.e., for rates of 10^{20} $[m^{-3}s^{-1}]$ or higher. Both the nucleation rate and the condensate mass fraction cannot be taken from our measurements. However, detailed quantitative comparison of experiments and numerical results has been made for the static pressure distribution and the global 2-D structure of the location and geometry of continuous compressions, shocks, and shock reflections.

Subcritical Heat Addition from Homogeneous Condensation — Nozzle No. 1

The first example is the recomputed uppermost schlieren picture in the first row of Figure 1. Figure 3 shows from top the nucleation rate, the condensate mass fraction, and the static pressure disturbances; flow direction is from the left. The nucleation process concentrates approximately symmetric around the transonic throat region. It rises in about 70 μs, respectively 20 mm in distance to the maximum level of 10^{25} $[m^{-3}s^{-1}]$ and it quickly decreases to zero with the growth of the condensate mass fraction g in the right hand supersonic part of the nozzle. The nucleation rate and the onset of the condensate mass fraction follow approximately parabola-like isolines of the density and temperature. So far, the variation of the gradients normal to the streamlines is still small and the nonequilibrium process develops 1-D in this global 2-D flow field. The isolines of the condensate mass fraction behind the onset and the static pressure plot show clearly the

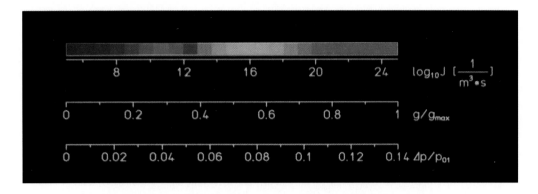

Figure 2   Color chart for the numerical results of the homogeneous nucleation rate J per unit time and volume, the condensate mass fraction g/g_{max} and the diabatic static pressure disturbance $\Delta p/p_{01}$ in comparison with the adiabatic flow.

existence of an "x-shock". It reduces the supersaturation turning back toward equilibrium and delays the growth of the condensate a little near the plane of symmetry of the nozzle.

Although the nucleation process starts in subsonic flow upstream of the throat, significant droplet formation and the correlated compression are definitely located in the right hand supersonic part. This allows the modeling up to the condensation onset as isentropic, respectively as frozen flow.

Supercritical Heat Addition from Homogeneous Condensation — Nozzle No. 1

Our second numerical example (Figure 4) reproduces the lowest picture of the first column in Figure 1. We recognize in flow direction two separated, well-structured nucleation zones. The first nucleation process starts upstream in subsonic flow and it abruptly breaks down across the straight shock. The strength of this shock and with that the maximum supersaturation decreases continuously from the center line toward the nozzle wall. Both the lower nucleation rate and the lower absolute number of nuclei formed near the plane of symmetry just upstream of the shock initiate the continuation of the nucleation process in subsonic flow behind the shock. Generally speaking, the shock causes an additional delay toward the equilibrium state. The complex 2-D structure of this interaction at the microscopic scale, and of the process as a whole, is manifested in the striking wavy structure of the second nucleation zone and the deformation of the constant condensate mass fraction contours in Figure 4. Again, this is not in contrast to the formation of a straight shock, rather it emphasizes the singular situation of this supercritical configuration.

The supersonic outflow of this nozzle consists, aside of the carrier gas, of growing droplets from nuclei of the first nucleation zone and of nuclei of approximately the same concentration produced in the supersonic expansion behind the first nonequilibrium condensation zone. The low reservoir vapor pressure of the mixture prevents further growth of these nuclei to droplets.

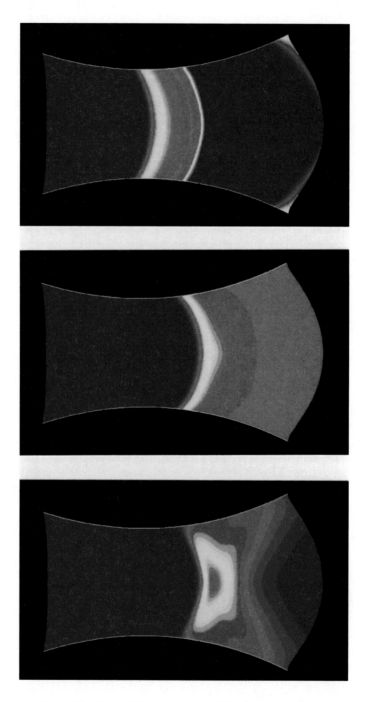

Figure 3 Recomputation of the first schlieren picture of the first row of Figure 1. Nozzle No. 1 "x-shock" and completely subcritical heat addition in supersonic flow. From top: Nucleation rate, condensate mass fraction, and static pressure disturbance in comparison with the adiabatic flow. Reservoir conditions: $\phi_0 = 37.5\%$, $p_{01} = 1\text{bar}$, $T_{01} = 303.9\text{K}$.

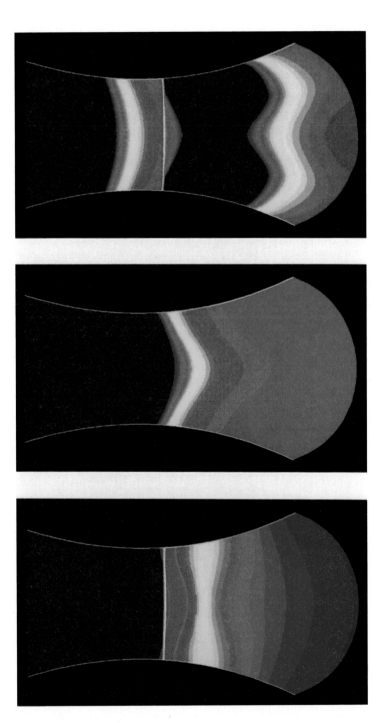

Figure 4   Recomputation of the first schlieren picture of the third row of Figure 1. Nozzle No. 1 complete supercritical heat addition in the stationary flow. From top: nucleation rate, condensate mass fraction, and static pressure disturbance in comparison with the adiabatic flow. Reservoir conditions: ϕ_0 = 66.5%, p_{01} = 1bar, T_{01} = 292.8K.

This supercritical example is just a limiting case at the onset of nonstationary flow. Within the discretization of the colored representation nearly no condensate, resp. no heat is added to flow upstream of the shock. The Mach number just in front of the shock is less than 1.1. On the other hand, the known natural sensitivity of transonic flow to any small parameter variation allows an effective quantitative control of our numerical results by comparison with experimental flow visualizations.

Mixed Sub- and Supercritical Heat Addition from Homogeneous Condensation — Nozzle No. 2

Our previous examples were either fully subcritical or completely supercritical flows. In between we expect a combination of both depending on the individual streamline of the 2-D flow field. For a better demonstration we take the second picture in the second row of Figure 1. Obviously there is a normal shock and a lenticular local subsonic area, but it is not extended to the nozzle walls (Figure 5). Outside, due to the streamline curvature, the heat addition develops subcritical in pure supersonic flow. From the condensate mass fraction plot it follows that along the center line the flow immediately becomes supercritical. At the same time the supersonic flow compresses weakly along the wall streamlines.

This behavior is a consequence of the sensitive interaction of the gradients of the heat addition and the local gradients of the flow variables along the actual streamline. It demonstrates further the structure which we have to expect at the onset of periodic oscillations in 2-D flows. The flow becomes nonstationary at first near the plane of symmetry of the nozzle (Schnerr 1986).

Conclusions

The presented work is the basis for the understanding and correct interpretation of more complex interactions of transonic flows and nonequilibrium condensation processes over airfoils and in cascades. In multidimensional flows 1-D modeling of the phase transition process near the onset is possible. In flows with diabatic structures, dominated by so-called subsonic heating fronts, pressure measurements and interferometry lead to a misinterpretation of the signals at the condensation onset. The relevant parameter of Laval nozzles with respect to this behavior is the nondimensional wall curvature at the throat y^*/R^*, whereas $\sqrt{y^*R^*}$ is the characteristic length for the cooling rate of the nozzle flow. Further numerical work is necessary for corresponding comparisons of experimental and numerical flow visualizations in the nonstationary flow regime.

Acknowledgments

This work was partially supported by the Deutsche Forschungsgemeinschaft (DFG contract Zi 18/31) and the Klein, Schanzlin and Becker Stiftung (KSB contract 1111). The authors are grateful to G. Mundinger for the preparation of the colored prints of the numerical results, and we would like to express our thanks to Peter P. Wegener of Yale University for many valuable suggestions and helpful discussions during the research on this subject.

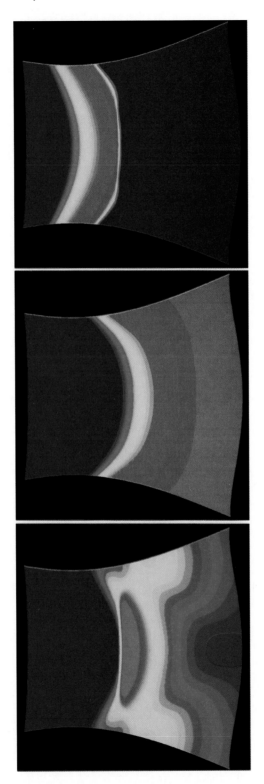

Figure 5   Recomputation of the second schlieren picture of the second row of Figure 1. Nozzle No. 2 mixed sub- and supercritical heat addition. From top: nucleation rate, condensate mass fraction, and static pressure disturbance in comparison with the adiabatic flow. Reservoir conditions: $\phi_0 = 59.5\%$, $p_{01} = 1\text{bar}$, $T_{01} = 292.7\text{K}$.

References

Bartlmä, F.: Ebene überschallströmung mit relaxation, in: *Applied Mechanics* (Ed. Görtler, H.), pp. 1056–1060. Proc. IUTAM Symp. München, RG, 1964. Berlin, Heidelberg, New York: Springer (1966).

Bartlmä, F.: *Gasdynamik der Verbrennung.* Springer-Verlag, Wien, New York (1975).

Bratos, M., Meier, G. E. A.: Two-dimensional, two-phase flows in a Laval nozzle with nonequilibrium phase transition. *Arch. Mech.,* Vol. 28, pp. 1025–1037 (1976).

Clarke, J. H., Delale, C. F.: Nozzle flows with nonequilibrium condensation. *Phys. Fluids,* Vol. 29, pp. 1398–1413 (1986a).

Clarke, J. H., Delale, C. F.: Supercritical shocks in nozzle flows with nonequilibrium condensation. *Phys. Fluids,* Vol. 29, pp. 1414–1418 (1986b).

Delale, C. F.: An asymptotic predictive method for gasdynamics with nonequilibrium condensation, in: *Proc. IUTAM Symposium Adiabatic Waves in Liquid Vapor Systems* (Eds. Meier, G. E. A., Thompson, P. A.) Springer-Verlag, Berlin, pp. 143–157 (1990).

Dohrmann, U.: Ein numerisches verfahren zur berechnung stationärer transsonischer strömungen mit energiezufuhr durch homogene kondensation. *Dissertation,* Universität (TH) Karlsruhe, Fakultät für Maschinenbau (1989).

Guha, A., Young, J. B.: Stationary and moving normal shock waves in wet steam, in: *Proc. IUTAM Symposium Adiabatic Waves in Liquid Vapor Systems* (Eds. Meier, G. E. A., Thompson, P. A.), Springer-Verlag Berlin, pp. 159–170 (1990).

Hall, R. M.: Onset of condensation effects as detected by total pressure probes in the Langley 0.3-meter transonic cryogenic tunnel. *NASA Tech. Memo.* 80072 (1979).

Hermann, R.: Der kondensationsstoß in Überschall-Windkanaldüsen. *Luftfahrtforschung,* Vol. 19, pp. 201–209 (1942).

Kilgore, R. A.: Special course on advances in cryogenic wind tunnel technology. *AGARD Report No. 774,* von Kármán Institute, Rhode-Saint-Genèse, Belgium (1989).

Moheban, M., Young, J. B.: A time-marching method for the calculation of blade to blade non-equilibrium wet steam flows in turbine cascades. *Inst. Mech. Eng. Conf. Publ.,* Computational Methods for Turbomachinery, Paper C 76184, Birmingham (1984).

Oswatitsch, K.: Die nebelbildung in windkanälen und ihr einfluß auf modellversuche. *Jahrb. Dtsch. Luftfahrtforsch.,* Vol. 1, pp. 692–703 (1941).

Prandtl, L.: Allgemeine überlegungen über strömungen zusammendrückbarer flüssigkeiten, in: *Reale Accademia d'Italia,* Fondazone Alessandro Volta. Atti dei Convegni 5. Le Alte Velocita in Aviazone. September 30 - October 6, 1935 - XIII, Roma. 1st ed. 1936 - XIV.

Schnerr, G. H.: Homogene kondensation in stationären transsonischen strömungen durch lavaldüsen und um profile. *Habilitationsschrift,* Universität (TH) Karlsruhe, Fakultät für Maschinenbau (1986).

Schnerr, G. H.: 2-D transonic flow with energy supply by homogeneous condensation: onset condition and 2-D structure of steady laval nozzle flow. *Exp. Fluids,* Vol. 7, pp. 145–156 (1989).

Schnerr, G. H., Dohrmann, U.: Theoretical and experimental investigation of 2-D diabatic transonic and supersonic flow fields, in: *Proc. IUTAM Symposium Transsonicum III,* Göttingen, May 24-27, 1988 (Eds, Zierep, J., Oertel, H.), Springer-Verlag Berlin, pp. 131–140 (1989).

Schnerr, G. H., Dohrmann, U.: Numerical investigation of nitrogen condensation in 2-D transonic flows in cryogenic wind tunnels, in: *Proc. IUTAM Symposium Adiabatic Waves in Liquid Vapor Systems* (Eds: Meier, G. E. A., Thompson, P. A.), Springer-Verlag Berlin, pp. 171–180 (1990a).

Schnerr, G. H., Dohrmann, U.: Transonic flow around airfoils with relaxation and energy supply by homogeneous condensation. *AIAA-J.,* Vol. 28, pp. 1187–1193 (1990b).

Schnerr, G. H., Dohrmann, U.: Drag and lift in nonadiabatic transonic flow. *AIAA-paper* 91-1716. AIAA 22nd Fluid Dynamics, Plasma Dynamics & Lasers Conference, June 24-26, 1991, Honolulu, Hawaii.

Schnerr, G. H., Bohning, R., Breitling, T., Jantzen, H.-A.: Compressible turbulent boundary layers with heat addition by homogeneous condensation. *AIAA-J.,* Vol. 30 (3), pp. 1284–1289, (1992).

Volmer, M., Weber, A.: Keimbildung in übersättigten gebilden. *Z. Phys. Chemie,* Vol. 119, pp. 277 (1926).

Volmer, M.: *Kinetik der Phasenbildung.* Steinkopff, Leipzig (1939).

Wegener, P. P.: Condensation phenomena in nozzles. *Progr. Astronaut. Aeronaut.,* Vol. 15, pp. 701–724 (1964).

Wegener, P. P.: Nucleation of nitrogen: experiment and theory. *J. Phys. Chem.,* Vol. 91, pp. 2479–2481 (1987).

Wegener, P. P.: Cryogenic wind tunnels and the condensation of nitrogen. *Exp. Fluids,* Vol. 11, pp. 333–338 (1991).

Young, J. B., Guha, A.: Normal shock-wave structure in two-phase vapour-droplet flows. *J. Fluid Mech.,* Vol. 228, pp. 243–274 (1991).

Zierep, J., Lin, S.: Bestimmung des kondensationsbeginns bei der entspannung feuchter Luft in überschalldüsen. *Forsch. Ingenieurwesen,* Vol. 33, pp. 169–172 (1967).

Zierep, J., Lin, S.: Ein ähnlichkeitsgesetz für instationäre kondensations vorgänge in der laval-düse. *Forsch. Ingenieurwesen.* Vol. 34, pp. 97–99 (1968).

Zierep, J.: Similarity Laws and Modeling, in: *Gasdynamics Series of Monographs* (Ed: Wegener, P. P.), Marcel Dekker, New York (1971).

Zierep, J.: Theory of flows in compressible media with heat addition. *AGARDograph* No. 191 (1974).

Zierep, J.: Theory of flows in compressible media with heat addition. *Fluid Dyn. Trans.,* Vol. 10, pp. 213–240, Warszawa (1980).

Zierep, J.: *Strömungen mit energiezufuhr.* G. Braun Verlag, Karlsruhe, 2nd ed. (1990).

chapter ten

Three-Dimensional Visualization of Medical Images

Yasuzo Suto

Department of Information and Communication Technology
School of High-Technology for Human Welfare
Tokai University, Shizuoka, Japan

Abstract — *Diagnostic imaging technologies have undergone remarkable development, making it possible to acquire high-precision fluoroscopic images (digital x-ray images), to obtain high-contrast CT and to reconstruct slice in any desired sectional plane using magnetic resonance (MR) images. The use of such imaging technologies has greatly contributed to diagnostic accuracy. However, in general, three-dimensional (3-D) visualization has been found to be more effective in the evaluation of bones or organs with complex structures. Most medical imaging modalities have been limited to two-dimensional (2-D) visualization due to restrictions in computer processing speed and memory capacity as well as high cost.*

Recently significant technological advances, such as the improvement in semiconductor technology, have led to rapid progress in medical electronics, and further research into the practical applications of 3-D visualization techniques is anticipated.

In this paper, I will discuss the 3-D visualization methods and clinical applications of 3-D visualization.

Introduction

In general, 3-D visualization has been found to be more effective in the evaluation of bones or organs with complex structures. Multislice x-ray CT and magnetic resonance images (MRI) and stereoscopic digital x-ray images inherently contain 3-D data. However, most medical imaging modalities have been limited to 2-D visualization due to restrictions in computer processing speed and memory capacity, as well as high cost.

Recently significant technological advances, such as the improvement in semiconductor technology, have led to rapid progress in medical electronics, and further research into the practical applications of 3-D visualization techniques is anticipated. Although 3-D visualization has a wide range of potential applications, it is basically applicable to the fields of diagnosis and therapy.

At present, physicians rely on their clinical experience to mentally visualize 3-D images by viewing images slice-by-slice or by observing fluoroscopic images. If this mental process could be automated, if would be of great value in realizing objective image evaluation.

Furthermore, 3-D visualization has been found to be effective in the precise assessment of the location and structure of lesions and their surrounding structures, both before and after therapeutic intervention.

In this paper, I will discuss the 3-D visualization methods and clinical applications of 3-D visualization.

Characteristics of Medical Images

Most clinical images are 2-D and are generated by measuring the changes in energy in a data carrier, such as an x-ray or ultrasonic beam, which has passed through the patient's body. These 2-D images can be divided into two main types: tomographic (slice) images such as those obtained using CT or MRI, and projection images such as those obtained by conventional x-ray examinations. In order to reconstruct 3-D images, these tomographic or projection images must contain 3-D anatomical data. Such 3-D data can be obtained by piling up a series of tomographic images (multislicing) or by obtaining projection data from a number of different directions (such as in stereoscopic imaging) (Figure 1).

The data containing 3-D anatomical information are then stored in computer memory, and 3-D images are reconstructed using a 3-D visualization algorithm, described in the next section.

3-D Visualization Algorithms

A variety of methods are employed for 3-D visualization, the most common being the surface visualization method, which permits stereoscopic image reconstruction of the surface structure of tumors, bones, blood vessels, and organs such as the brain with its complex surface anatomy. Recently, however, there has been increasing interest in methods for visualizing information concerning internal structures as well as the external surface. Such methods are called *volume visualization*. Table 1 shows an outline of volume visualization algorithms, which can be roughly divided into two types; one type to visualize the surfaces of objects, and another to visualize the internal structures of objects.

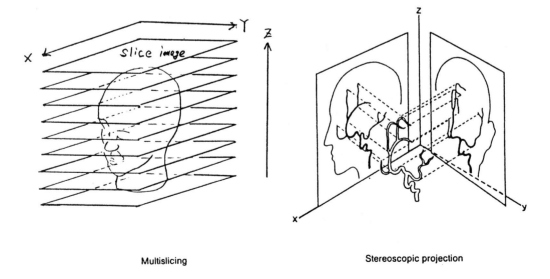

Multislicing   Stereoscopic projection

Figure 1   Three-dimensional data structure of medical images.

Table 1 Volume Visualization Algorithm

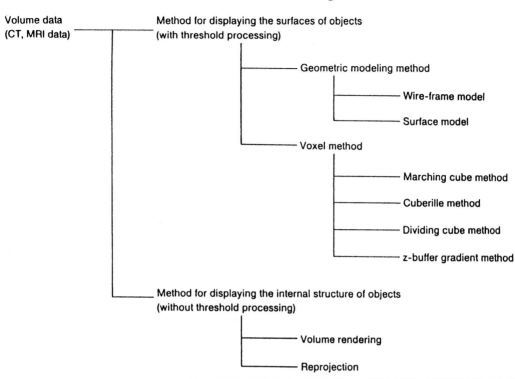

Methods for Visualizing the Surfaces of Objects

A variety of different techniques can be used to visualize the surface of objects, such as the geometric modeling, voxel, and stereoscopic methods. Threshold processing is employed in all these methods. The general processing procedures for these methods are shown in Figure 2.

First, using CT or MRI slice, or stereoscopic images acquired using digital subtraction angiography (DSA), contour extraction is performed for the region of interest. If the blood vessels are to be examined, core extraction is performed. Then, a threshold value is specified for conversion of the data to a binary image. Slices of this binary image are then used to generate 3-D images. The direction of the viewpoint is specified, the image is projected onto the z-buffer coordinate axis, and shading is performed. Three methods used to generate 3-D images are described in the following sections.

Geometric modeling method

From multislice images, the contours of the body surface, organs, or tumors are extracted as line images, which can then be converted into two types of 3-D images: wire-frame models generated using interslice mesh combination (tiling), and surface models generated by painting the mesh. First, a pair of slice contours are divided at arbitrary intervals. After the data have been generated for a series of points, pairs of points are connected with straight lines and a wire-frame image is generated consisting of triangular or square patches (Figure 3).

In order to convert a wire-frame image into a surface model image, the patches are shaded using a shadow factor determined by the angle θ between the normal line of the patch and the direction of the line source (or the direction of the viewpoint) (Figure 4). The shadow factor 1 of the patches calculated from the normal vector N, the unit vector in the

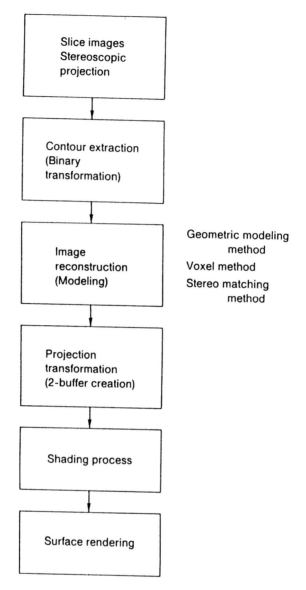

Figure 2   General processing procedures for displaying surfaces of objects.

direction of the light source L, and the angle θ between the normal line and the light source is obtained from the following:

$$I = C_1 (N \cdot L) + C_2 = C_1 \cos \theta + C_2$$

where C_1 and C_2 are arbitrary constants.

Figure 5 shows a composite 3-D image of a wire-frame image of the scalp, a surface image of the cerebral surface.

Voxel method

In this method, voxel (volume element) is first defined as 3-D pixels (cubes) containing 3-D information, and a 3-D imaging space (multislice imaging space) is created using these voxels (Figure 6). In this case, interpolation is required to balance the spatial resolution (in the x and y axis) and the slice resolution (interslice distance, z axis) so that each voxel is

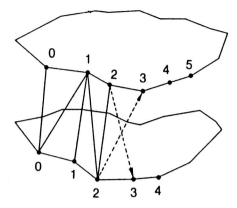

Figure 3 Construction method of patches.

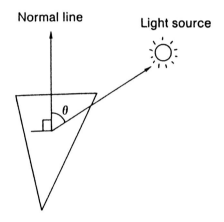

Figure 4 Shading method of patches.

an exact cube. The numerical values assigned to voxels are usually CT values in CT images and proton density values, T_1 or T_2 in MR images. The process of 3-D image reconstruction using the voxel method is described below.

1. **Extraction of the region of interest:** first, threshold processing is performed to obtain a 3-D bone or tumor image in the 3-D imaging space (Figure 6). For example, a threshold value of 40 is assigned to each voxel and the voxels containing a number higher than 40 are automatically extracted and, as a result, a binary voxel space containing 0 and 1 values is generated (Figure 7).
2. **Projection of 3-D images:** after the binary voxel space has been generated, the viewpoint is set for transforming and projecting the data onto the z-buffer coordinate system and a z-buffer (depth) image is generated. The two major z-buffer transformation methods are described below:
   • **Back-to-front method:** voxels with a value of 1 in the binary voxel space (Figure 7) undergo 3-D projection transformation in order, starting with those farthest from the viewpoint, and the z-buffer image is generated. The back-to-front method eliminates those voxels farther from the viewpoint (hidden surface processing), and generates a z-buffer image consisting of the voxels near the viewpoint.

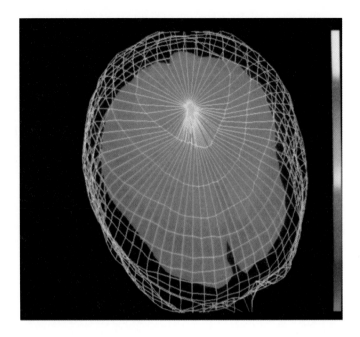

Figure 5   Brain 3-D image constructed from MR images using the geometric modeling method.

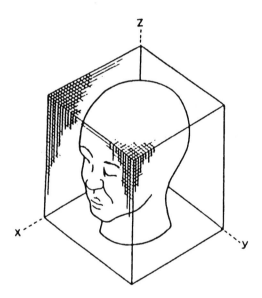

Figure 6   Creation of the voxel space.

- **Ray tracing:** rays are emitted from the viewpoint, and the distance (depth values) to the objects (voxels) are obtained and used as z-buffer pixel values. The hidden-surface-processed z-buffer image (depth value image) is automatically obtained in this method.
3. **Shading:** the hidden-surface-processed z-buffer image itself does not provide sufficient contrast, therefore shading must be performed. Although several methods can be used for shading, gradient shading is used most frequently. Figure 8 (left) shows a 3-D image in which only the depth (z-buffer) values were used for shading,

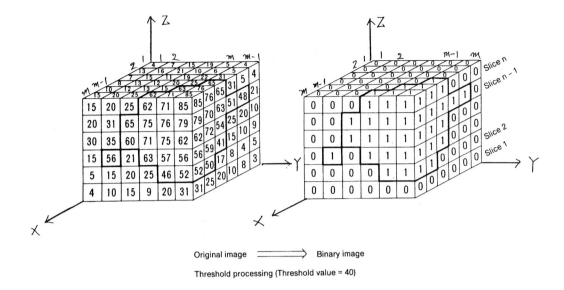

Original image ══════⟹ Binary image

Threshold processing (Threshold value = 40)

Figure 7 Thresholding of the voxel space.

while Figure 8 (right) shows the same image with gradient shading applied to the depth values.

4. **Colored, semitransparent display:** opacities are assigned to several 3-D images where z-buffer shading has been applied, then color processing is performed to generate colored, semitransparent volume image (Figure 9). The images shown in Figure 9 are color 3-D images obtained using threshold processing for the brain surface (green), a brain tumor (orange), and the falx cerebri (blue). The tumor (orange) is visible through the semitransparent falx cerebri (blue).

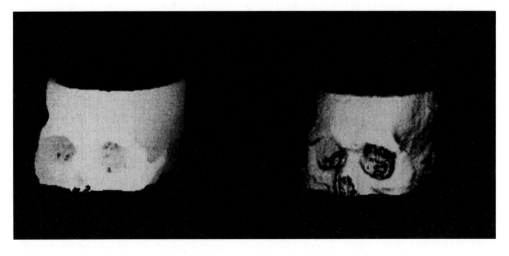

Figure 8 3-D image of the skull generated using the voxel method. The depth value was used for the image on the left and the gradient method was used for shading processing for the image on the right.

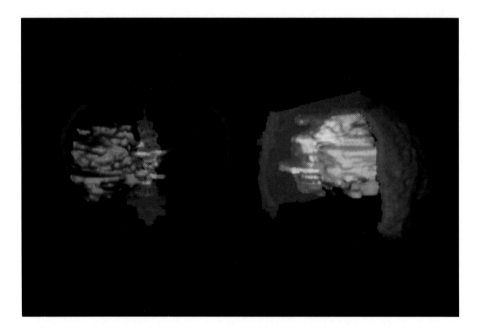

Figure 9   Color composite 3-D image generated from x-ray CT images.

Stereo matching method

Clinical images with inherent 3-D data structure are divided mainly into multislice images such as CT and MR images, and projection images such as conventional x-ray images. Projection images are based on the integrated radiation dose received from a certain direction after the radiation has passed through the patient's body. Therefore, the data does not contain 3-D information. However, as shown in Figure 1 (right), stereoscopic DSA allows images to be obtained from two different directions, and 3-D images can be obtained by calculating the depth (z-coordinate value) from the parallax between the two projected images. The coordinate value (x, y, z) in the reconstructed 3-D space, corresponding to the points PR and PL on the two projected images, are obtained as points of intersection between the OL-PL line segment and the OR-PR line segment (Figure 10). l_1, l_2, and θ are already known, while dL and dR are calculated from the images after the pixel size has been determined. The angle θ can be obtained from the parallax (8 to 10°) in stereoscopic DSA, but can be also defined as an arbitrary angle.

A stereoscopic 3-D DSA image of the cerebral vessels is shown in Figure 11.

Methods for Displaying the Internal Structure of Objects

One method for displaying the internal structure of objects is called volume rendering, and has generated interest as a new visualization technique for 3-D objects including the human body. Here, "volume" indicates continuous functional data concerning not only the surface anatomy but also the internal characteristics (such as temperature or density distribution), or discrete data such as observed values. Opacity and RGB color data are assigned to each voxel of the surface and the internal structure is visualized simultaneously using ray casting. In other words, volume rendering differs from conventional methods which employ threshold processing, such as binary classification, and visualize the surface only. In volume rendering, anatomical structure is classified based on the theory of statistical optimization and an opacity is assigned to each voxel. Therefore, anatomical structure can be visualized in a more authentic, detailed, and accurate manner.

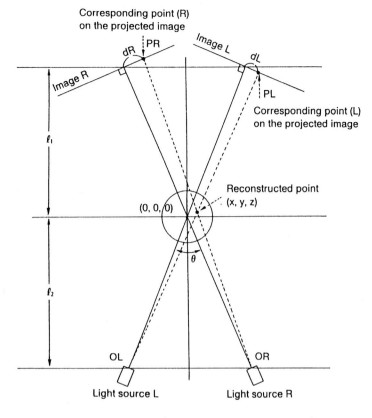

Figure 10 Principle of stereoscopic image generation.

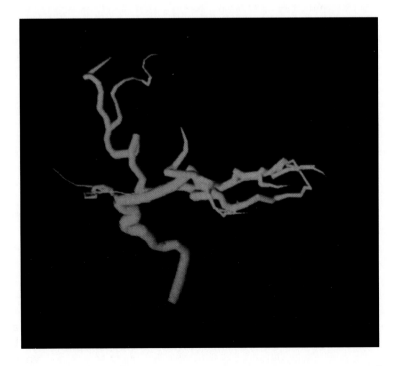

Figure 11 3-D image of the cerebral vascular obtained using stereo matching.

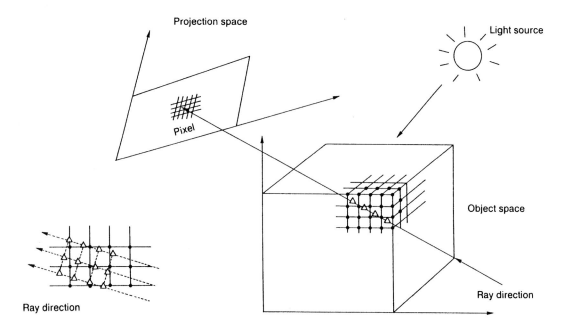

Figure 12   Principle of volume rendering.

In addition, the methods used to assign a color and opacity to each voxel permit the visualization of a variety of biological parameters to meet the specific diagnostic application.

In volume rendering, multislice images handled as volume data are first divided into voxels, and a 3-D array is created. Then, as shown in Figure 12, a 3-D model defined in 3-D space is subjected to voxel tracing (ray casting) — in practice this is performed from front to back, but this explanation describes tracing from back to front — and not only boundary surface of the structures, but also internal information, can be visualized as multiple transparent layers.

The general process employed in volume rendering is shown in Figure 13. As shown in this chart, the color intensity for each voxel is obtained by shading during ray casting and the opacity is determined by classification. The color intensity and opacity are then combined to generate a rendered image. In this case, since rays generally pass through voxels at arbitrary angles, changing the arrangement of voxel lattice points (resampling) is required in each step of ray casting. Interpolation is performed for $C_k(x_i)$ and $\alpha(x_i)$, which are assigned to a nearby voxel, and the color intensity $C_k(u_i)$ and opacity $\alpha(u_i)$ are obtained (u_i indicates the resampling coordinate). Next, these two values are combined for rendering. Opacities are assigned to voxels located along the line in the direction of the viewpoint, and the incident ray $C_{in,k}(u_i)$ and inherent illumination $C_k(u_i)\,\alpha(u_i)$ are combined and transmitted in succession to each voxel nearer the viewpoint, and the light passing out of the last voxel, when the opacity is 0, $C_{out,k}(u_i)$, is used as the pixel value for rendering. Figure 14 (left) shows rendering simulation applied to brain CT images for the observation of the skull through the transparent scalp with the opacity of the scalp set to 0.3. The same processing was performed with the opacity of the scalp increased to 0.7 and to 1.0, and the results are shown in Figure 14 (center and right). Figure 15 shows 3-D images of the brain obtained from MR images using volume rendering.

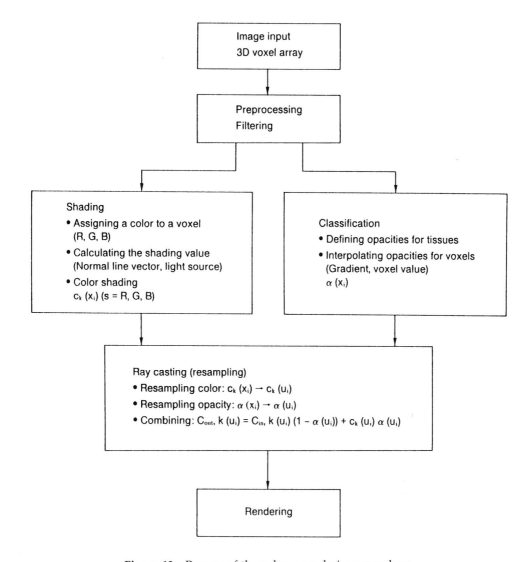

Figure 13   Process of the volume rendering procedure.

The structures of blood vessels and the cerebrum are clearly demonstrated. Unlike the synthesized color images shown in Figure 9, the elimination of threshold processing permits visualization of information regarding the internal structure of the brain without signal loss and in a more natural manner.

Clinical Applications of 3-D Display Techniques

Three-dimensional imaging has been found to be effective in a variety of fields, particularly diagnostic imaging and surgical simulation. First, in the field of diagnostic imaging, 3-D visualization is extremely helpful for clarifying the precise 3-D structure of tumors or skeletal abnormalities and complements the diagnostic information obtained using conventional 2-D imaging. Figure 16 shows 3-D images of scoliosis generated from CT images; the spinal deformity can be clearly observed. Furthermore, the ability to visualize individual

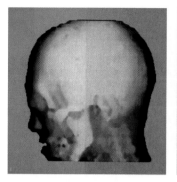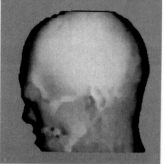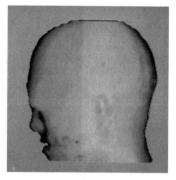

Figure 14   Volume rendering applied to brain x-ray CT images, with the opacity changed from 0.3 (left), to 0.7 (center), to 1.0 (right).

vertebral bodies separately from the spinal column improves diagnostic accuracy and permits the volume of each vertebral body to be calculated. Such images cannot be obtained using 2-D imaging.

Next, 3-D imaging is effective in surgical simulation (in planning the surgical approach using 3-D images generated from CT and MR image data before the procedure, measuring the size of the tumor, and performing surgical planning such as cutting and repositioning) and contributes greatly to maximizing the safety of surgical procedures.

Figure 17 shows a cutting and repositioning simulation performed for a case of ocular hypertelorism, a congenital craniofacial deformity. The bones surrounding the orbit were divided into three sections (Figure 17a). The original 2-D slice image was used to determine the cutting depth at a main point on the cutting path (Figure 17b). The central section was then removed (Figure 17c) and the right and left sections repositioned (Figure 17d), completing the simulation.

Conclusion

The algorithm for 3-D visualization and the clinical applications of the 3-D images generated using these algorithms have been discussed in this paper. The amount of data required for 3-D visualization is enormous and the number of calculations can be astronomical. Moreover, recent medical images have achieved a high level of precision and provide multiple-gradient information, making fast processing capabilities even more essential. Therefore, there is increasing demand for the use of extremely large computers or supercomputers. On the other hand, rapid advances in workstation performance have recently been achieved, leading to the greater availability of systems offering superior cost/performance. Thus, 3-D visualization can be expected to become routine in the near future.

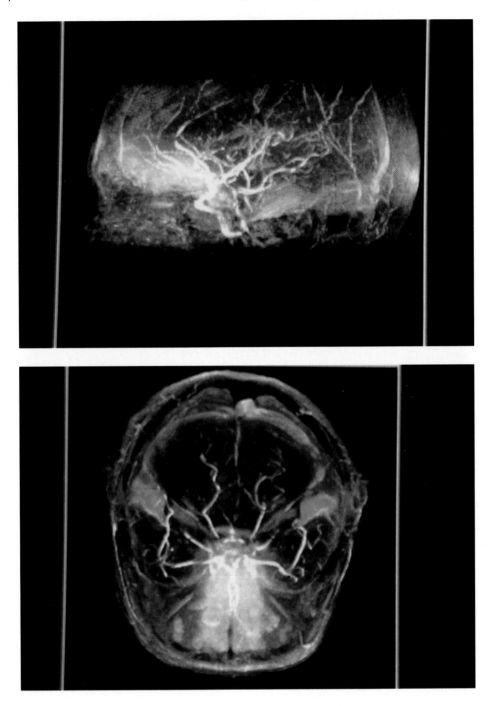

Figure 15 Volume rendering applied to brain MR images. Lateral view (top) and transverse view (bottom).

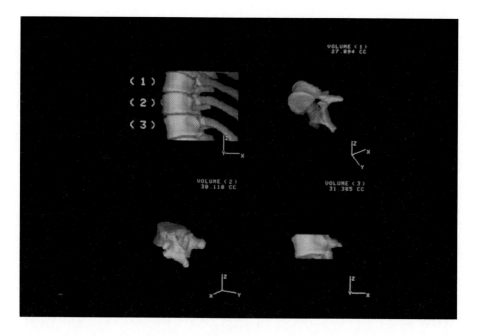

Figure 16   3-D images of scoliosis.

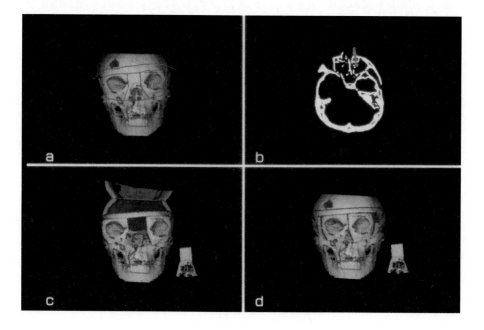

Figure 17   Surgical simulation applied to ocular hypertelorism: a. drawing cutting paths; b. specifying the cutting depth based on 2-D slice images; c. cutting and removing the central bone section; d. repositioning bone sections.

References

Suto, Y., 1990: 3-D display, *Electrophotography*, Vol. 29, Part 2, pp. 221–229, (in Japanese).

Yokoi, S., 1989: Display techniques for clinical 3-D images, *JJME*, Vol. 3, Part 8, pp. 11–17, (in Japanese).

Levoy, M., Hanrahan, P., and Kaufman, A., 1990: Tutorial on volume visualization algorithms, *Visualization '90*, San Francisco.

Suto, Y., Huan, T., Furuhata, K., Uchimo, M., and Kouno, T., 1988: Three-Dimensional image display of brain tumors using a modified voxel method, *Med. Imag. Tech.*, Vol. 6, Part 3, pp. 280–291, (in Japanese).

Iwai, S., Yachida, M., and Tsuji, S., 1985: 3-D reconstruction of coronary artery from cine-angiograms based on left ventricular model, *Trans. Inst. Electron. Inf. Commun. Eng.*, Vol. J68-D(2), pp. 137–144, (in Japanese).

Levoy, M., 1988: Display of surfaces from volume data, *IEEE Comput. Graphics Appl.*, Vol. 8, Part 3, pp. 29–37.

Drebine, R. A., Carpenter, L., and Hanrahan, P., 1988: Volume Rendering, *Comput. Graphics*, Vol. 22, Part 4, pp. 65–74.

Suto, Y., Furuhata, K., 1991: The clinical usefulness and technical problems for medical three-dimensional display based on volume rendering, *Med. Imag. Tech.*, Vol. 9, Part 3, pp. 257–258, (in Japanese).

Suto, Y., Furuhata, K., Kouno, T., Uchino, M., 1990: Surgical simulation system, *Innervision*, Vol. 5, Part 5, pp. 2–7, (in Japanese).

Suto, Y., 1991: Three-dimensional surgical simulation system using x-ray CT and MR images, *Med. Rev.*, Part 35, pp. 32–42, (in Japanese).

Suto, Y., Asahina, K., Furufata, K., Kobayashi, M., Mano, I., Korokawa, H., and Kojima, T., 1988: Clinical application of medical three-dimensional display to orthopedic surgery, Proc. 27th Conf., *Japan Soc. ME BE*, pp. 429, (in Japan).

chapter eleven

Some Promising Aspects of the Ultrasonic Imaging Technique

Hajime Yuasa*
Yasutaka Tamura†

*Akishima Laboratory, Mitsui Engineering and Shipbuilding Company, Ltd.
†Faculty of Engineering, Yamagata University

Abstract — The acoustic imaging technique is better than optical imaging with respect to water turbidity, range, and resolution. Furthermore the acoustic imaging technique tends to have higher resolution and shorter range than conventional sonar. This report reviews the principle of acoustic imaging techniques on newly developed sonars; encoding aperture using an inclined linear array, 2-ch M-sequence encoding array transducer system, and multiple shot 3-D holographic sonar using a set of orthogonalized modulating signals. These are discussed in comparison with existing synthetic aperture array systems; synthetic aperture side scan sonar and multi-beam scanning sonar.

Introduction

Unfortunately, the energy of an electro-magnetic wave which includes an optical wave can easily be reduced in the medium of water. Although deep ocean water has 6-15 m of optical visibility, most ocean water is much more turbid. Near-shore water typically has 1-6 m visibility wherever we actively work and observe underwater construction or marine ranching. When we disturb the mud and silt on the bottom, the optical visibility range is often decreased to less than 1 m.

Fortunately, the energy of an acoustic wave easily propagates in the medium of water in spite of the optical turbidity of the mud and silt. In general, the resolution capability of acoustic imaging is, however, seriously lower than optical imaging, because the wavelengths of sound for underwater imaging are much longer than optical wavelengths.

Compared with a conventional sonar system, the underwater acoustic imaging technique has the distinction of indicating the form or characteristics of a target, while conventional sonar indicates its location only, even if both have similar kinds of hardware instrumentations and physical properties. Thus, the acoustic imaging technique is better than optical imaging with respect to water turbidity, range, and resolution. Furthermore the acoustic imaging technique tends to have higher resolution and shorter range than conventional sonar.

This report reviews the principle of acoustic imaging techniques on newly developed sonars:

- Encoding aperture using an inclined linear array,
- 2-ch M-sequence encoding array transducer system, and
- Multiple shot 3-D holographic sonar using a set of orthogonalized modulating signals.

These are discussed in comparison with existing acoustic imaging systems:

- Synthetic aperature side scan sonar and
- Multi-beam scanning sonar.

Existing Acoustic Imaging Systems

Type of Acoustic Imaging Systems

For scanning or traversing a single transducer in any direction repeatedly within a plane with a certain distance above an objective target, nondestructive inspections have been applied, for instance bondings of hairlike gold wire to the terminals of an IC (integrated circuit). The acoustic image pattern of the IC cannot immediately be obtained because of the scanning procedure.

Using electronic spatial scanning, the applications of medical ultrasonic linear array becomes advantageous for obtaining images of diseased parts of the living body. These kinds of images are restricted in the sectional or lateral plane perpendicular to the linear array.

In this review the discussion will concentrate on the imaging systems that obtain acoustic images in the vertical or focal plane, similar to a photograph, but adding the lateral which is typically presented by conventional sonars. The acoustic imaging systems should be followed up to determine how to get an image similar to an optical image.

The types of acoustic imaging systems that use transducer array can basically be classified into two categories. One is called the beamformed or beam scanning system which performs spatial processing first and then detection. The other is based on the holographic approach which first produces an acoustic hologram and then performs the spatial processing, sometimes called reconstructing the hologram.

The beamformed method is used in conventional sonars. In this chapter two kinds of typical, conventional sonars are outlined to express the distinction from the "holographic".

Synthetic Aperture Side Scan Sonar

Transmitting ultrasonic beams at an angle gazing to the bottom, side scan sonars display a lateral plane image as shown in Figure 1. The image is generally produced by a single sweep beam or by multiple performed beams. In the case of the single beam, a single transducer is mechanically moved through the water, sweeping a polar coordinate plane. However, an advanced "synthetic aperture side scan sonar" consists of a linear array of transducers and transmits a wider beam. The array electronically produces spatial beams by coherently combining the echoes from several transmit cycles.

These types of sonars are able to obtain bottom profiles, however they take a long time to display the polar coordinate plane of the image and needed empirical skills to explain well as Sutton (1979) remarked. Figure 2 shows an image of a sunken vessel obtained by the synthetic aperture side-looking sonar.

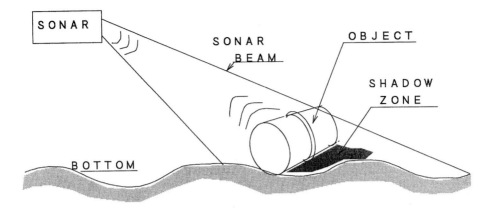

Figure 1   Synthetic aperture side-looking sonar. (From Sutton, J. L., *Proc. IEEE*, 1979.)

Figure 2   The acoustic image of a sunken vessel. (From Dowty Maritime Ocean Systems.)

Multibeam Scanning Sonar

As a synthetic aperture technique an electric company in Japan developed a multibeam scanning sonar Nitadori, Mano, and Kamata, 1980. This consists of a receiving array of 32×32 elements arranged in a square with 320 mm sides and a transmitting array of 4×4 elements with sides of about 1 m. Transmission frequency is 200 kHz, and the spatial coverage is a 40° field of view with measured resolution of 0.4°.

One image requires pulsing the transmitters 64 times. With one resolution cell from an observing field, the transmit patterns are sifted to focus with the phase of each pulse of transmitters. The receiving and processing electronics need high-level techniques to produce a 128×128 image displayed by a digital computer and an FFT processor on a TV monitor every 2 s.

This method, however, does not apply a pure holographic approach because of the definition of the classification mentioned previously.

New Concept of Acoustic Imaging Systems

Encoding Aperture Using an Inclined Linear Array

Basic concept

A new echolocation system named EAR (Encoding Array Receiver) has been studied and demonstrated by computer simulations and preliminary experiments (Tamura 1987 and Tamura and Akatsuka 1991). This requires no mechanical or electrical scanning to determine the 2-D cross section of acoustic images. Only one or two signal channels are needed to get the images which can be reconstructed from the data acquired by means of a single transmitting and receiving process.

Conventional systems

Conventional systems usually need an array of transducers and parallel electronic circuits to obtain reconstructed images. However, these parallel features complicate the systems and raise the cost.

Basic idea

This system consists of a coded and inclined array which include a transmitter set at the center point of the array of receiving transducers. Since a single signal channel is connected to an array of receiving transducers and the sensitivity of the transducers is modulated by means of an appropriate code sequence, received wavefronts are encoded into a serial signal. When a wide-band acoustic signal is transmitted, the array of receivers encodes any received wavefront into a serial signal with the accumulated echoes specified by the distance and the direction of the targets.

Theory of acoustic imaging

Figure 3 shows the schematic diagram of the EAR system. A linear array with wide-band receiving sensitivity has the angle of θ_A from the aperture plane watching the right direction. The interval of N transducers is arranged d. Figure 4 shows a scheme of received echoes with a train of pulses detected by the array. The interval of the pulses τ_T and the delay time of the pulse train τ_R are given by the following equations:

$$\tau_T = d \cdot \sin (\theta_A - \theta)/C \qquad (1)$$

$$\tau_R = 2R/C \qquad (2)$$

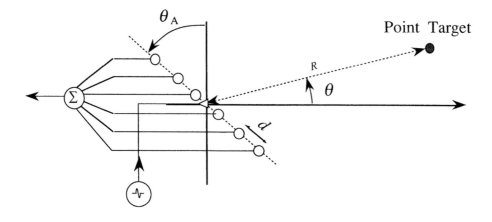

Figure 3   Schematic diagram of an EAR with a linear array.

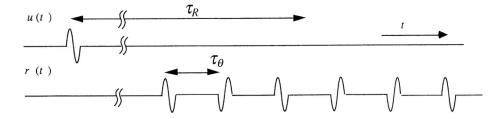

Figure 4 Transmitted pulse u(t) and waveform r(t) detected by an EAR.

where θ and R are the angle and the range of the target respectively. Accordingly, the direction θ and the distance R of the target can be determined from the interval and the delay of the pulse train.

Imaging with correlation

With a detected waveform r(t) and an image function s(x), the following equation can be given:

$$s(x) = r(t){\cdot}h(x,t)*dt \tag{3}$$

where h(x,t) denotes the detected waveform from a single point target located at a position x and * means the complex conjugation.

In the paraxial region of the array, an impulse response function h(x,t) is shown by the waveform function of transmitted acoustical pulse u(t) as follows:

$$h(x,t) = \sum_{i=1}^{N} m_i \cdot u\left(t - \frac{|x| + |x - x_i|}{C}\right) \tag{4}$$

where x_i is a position vector of the i-th receiver, m_i is a code value corresponding to the receiver, and C is the sound speed.

Array spatial modulation assigned by Golay code

For large range differences, spurious peaks tend to emerge in acoustic imaging. To elimi-nate the range side-lobes caused by the side-lobes of the auto-correlation function of the pulse-train h(x,t), the Golay code is employed on the spatial modulating sequence.

Golay code is a binary complementary code consisting of two codes, an A-code and a B-code. Figure 5 shows the alternatively assigned N receivers as the codes A and B of N/2 bits. By adding the auto-correlation function of the two codes, clearer images can be obtained without side-lobes.

Simulation

To demonstrate the above imaging method, a point spread function (PSF) was obtained by means of computer simulation. The simulation assumed a 2-D imaging system operating in mid-air. The angle of the inclination was 45°. An LFM (Linear Frequency Modulated) acoustical pulse was transmitted from the center of the array. The frequency band of the waveform was between 25 and 75 kHz.

Figure 6 shows the point spread function at the distance of 1 m. Figure 7 shows the relationship between the level of the spurious peaks and the number of receivers. The spurious peaks were evaluated as a ratio of the main peaks to the maximum spurious peak. A value about 11 dB was obtained when 16 receivers were used, and 16 dB for 64 receivers.

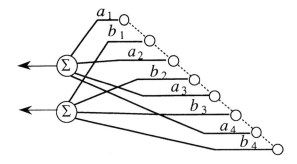

Figure 5   Assignment of codes to the elements.

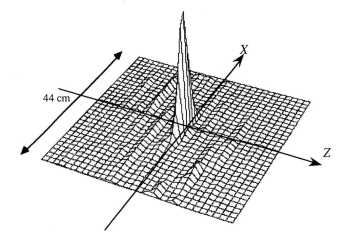

Figure 6   The point spread function of the imaging system.

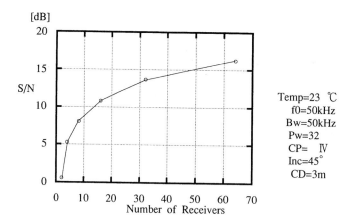

Figure 7   The level of the spurious peaks vs. the number of receivers.

Experiment

Photo 1 shows a view of the array. The experimental system has a pair of linear arrays configured symmetrically each of which consists of 16 microphones. The transmitters, designed to generate a nondirectional wavefront, are located in the center between the two

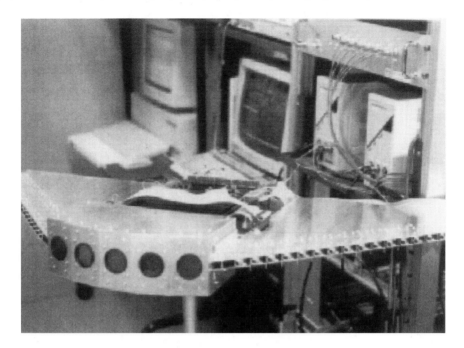

Photo 1 View of the array.

receiving arrays. The transmitted pulses with the center frequency 50 kHz are generated by means of a wave-generator. Transferring the detected wave forms digitized by A/D converters to an HP9000 workstation via a personal computer, the acoustic image can be reconstructed.

Preliminary experiment and simulation

Using the linear array on the right-hand side, an image of a wide-band sound source was reconstructed as a preliminary test. Among many spurious images, Photo 2a shows a bright spot which we recognize as the position of the sound source in a reconstructed cross-sectional image with 64 λ × 64 λ.

Photo 2b shows the image of a similar sound source reconstructed by computer simulation assuming the same array configuration as the experiment. The spurious peaks

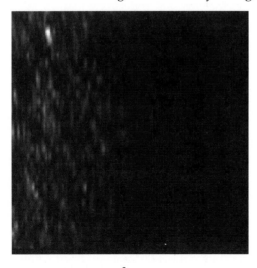

a   b

Photo 2 Reconstructed image: (a) experiment; (b) simulation, the imaging area is 64 λ × 64 λ (44 cm × 44 cm), the array is located at the left-hand side of the image.

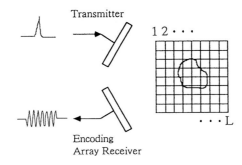

Figure 8   Diagram of the imaging system.

in the experiment are higher than in the simulation, probably because of an error in arranging the elements of the array or from a distortion of the received waveform.

M-Sequence Spatial Encoding Array Made of VDF-TrFE Copolymer

Principle of encoding array

If the transmitted waveform is a pulse of short duration and a single point target is being observed diagonally by the array, the output waveform of the array would become a train of pulses. From the respective interval and delays of a pulse train, the position of each target can be determined. In case of multiple targets the pulse trains overlap. However, if the pulses corresponding to distinct target positions are uncorrelated from each other, then a particular pulse train can be discerned and the targets spatially resolved.

Figure 8 shows, in diagrammatic form, the configuration for the observing a wavefront and reconstructing an image.

When wide-band ultrasonic waves are transmitted from a nondirectional tranducer, the ensuing reflected wavefronts can be detected by the array. With this arrangement, the function representing the reflectance map generated by an object and the actual waveforms can be characterized by the array. The function can be determined in discrete values by appropriate sampling.

A two-dimensional reflectance map represented by a vector s can be determined on L grid-points. Letting the detected waveform be r for the objects of reflectance map s is given by the following expression:

$$r = H \cdot s \tag{5}$$

Here, H is a transfer matrix composed of the impulse responses h_i for all subsequent grid-points $i = 1 \sim L$ which are the detected waveforms corresponding to each single point target at the grid points.

Image reconstruction

Employing a correlation operand in image reconstruction, a complex image of objects $\hat{s}$ is given as follows:

$$\hat{s} = H^* \cdot r \tag{6}$$

Here, a complex conjugated and transposed matrix H^* is modified by the echo vector r.

Figure 9 shows the result of a computer simulation of the 2-D point spread function of the imaging system. The point spread function for a j-th grid-point is given by:

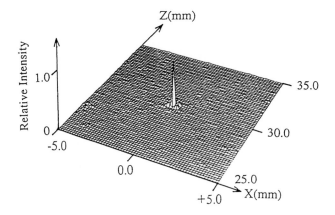

Figure 9 Two-dimensional point spread function for the 2-ch M-sequence encoding array transducer system shown by means of computer similation; a single wire target exists at X = 0 mm and Z = 30 mm.

$$P_j = H^* {\cdot} h_j \tag{7}$$

A target is assumed to exist at X = 0 mm and Z = 30 mm, where X is the parallel distance from the center of the transmitter and Z is the perpendicular distance from the transducer surface.

Spatial encoding array made of VDF-TrFE copolymer

Figure 10 shows a 2-ch M-sequence encoding array transducer system consisting of three films of VDF-TrFE copolymer with 60 μm thickness. The electromechanical coupling of the film is given by polarization after heat-treatment (Tamura and Akatsuka 1991).

A gold electrode is placed on one side of each of the films. Further, a flat gold electrode is then placed on the other side of the copolymer film acting as the transmitter and M-sequence pattern electrodes are then placed on the other side of the remaining two copolymer films. Later two transducers acting as receivers are placed symmetrically about the central transmitter. One side of the composite transducers is attached to a thin polyimide film of 25 μm thickness which is used as a matching layer. The other side is attached to ferrite rubber which acts as a backing material and ultrasonic absorbent. The center frequency of the transducers is 5 MHz.

Figure 11 shows the array consisting of an M-sequence electrode pattern with a length of 63 bits and binary values of +1 and −1. The width of each element in the array pattern is 0.15 mm and total length of the array pattern is 30.00 mm. Thus, the array transducer can

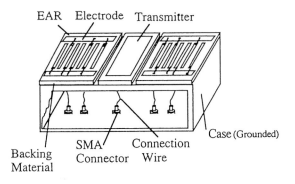

Figure 10 Configuration of the 2-ch M-sequence encoding array transducer system.

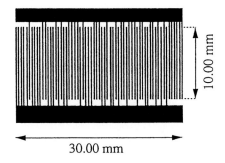

Figure 11   Electrode pattern of the M-sequence encoding array receiver.

receive a propagated ultrasonic wave with two different polarities encoded by the
M-sequence because the polymer transducer is active only in the area sandwiched between
the electrode patterns. The mechanical loss of the piezoelectrical polymer is large. Therefore the array transducers can be made simply by forming the electrode from the vapor
deposition of the films.

Figure 12 shows a block diagram of the experimental imaging system with the 2-ch
M-sequence encoding array transducer. A wide-band ultrasonic wave is radially generated
from the transmitter, then the reflected pulses from a target are received by the pair of
M-sequence encoding array receivers. In this experiment a piece of brass wire with diameter of 1.4 mm was used as a target.

The difference between the two signal amplitudes detected by each M-sequence encoding array receiver was calculated using a differential amplifier, and stored temporarily
in a digital memory (LeCroy 9400). Sending the data to a mini-computer (HP 9000/370),
the image of the target was reconstructed.

Figure 13 shows a sample of the received waveforms from a single target. Wave (a)
shows the wave pattern generated by computer similation, and (b) shows the wave pattern
received in the experiment. The simulated waveform is calculated by computer based on
the assumption that a cycle of sinusoidal pulse at 5 MHz is being received by an individual
electrode on the array receivers. Each wave form is digitized into 2048 points at a sampling

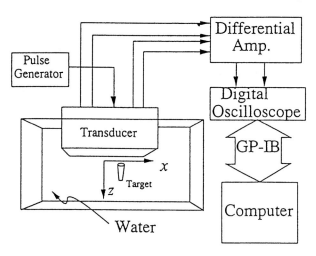

Figure 12   Block diagram of imaging system with the 2-ch M-sequence encoding array transducers.

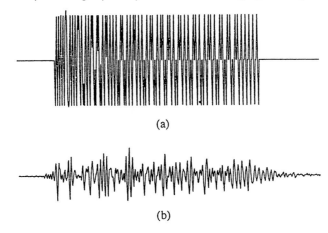

(a)

(b)

Figure 13   Ultrasonic reflected echo waveform from a single target using the 2-ch M-sequence encoding array tranducer system: (a) as simulated by computer; (b) as obtained in the experiment.

rate of 100 MHz. By comparing the experiment results with the simulated results, the former seems to be distorted. It appears that the distortion is caused by a distortion of the pulse that is detected by each electrode on the array. To suppress this distortion, the received waveforms are filtered using the transfer function H (ω) defined by:

$$H\ (\omega) = O\ (\omega)/I\ (\omega) \qquad\qquad (8)$$

where I (ω) is the Fourier transform of the waveform received by only one electrode on the array transducers and O (ω) is a cycle of sinusoidal pulses at 5 MHz. To obtain the objective waveform m(t), the following calculation is applied:

$$m(t) = F^{-1}\ [H\ (\omega)\ A\ (\omega)] \qquad\qquad (9)$$

where F^{-1} [H (ω) A (ω)] is the inverse Fourier transform, and A (ω) is the Fourier transform of the waveform actually detected by the M-sequence encoding array transducer system. Thus, the images are reconstructed after filtering the received waveforms.

Results

Figure 14 shows the image functions of the before filtering (a) and the after filtering (b). The latter indicates that the filtering process makes the sub-peak level lower.

Photo 3 shows the image of Figure 14(b) on a gray-scaled picture which displays a single wire as a target in the imaging area 10 mm × 10mm. The number of pixels is 512 × 512 and the imaging process takes about 10 min. The bright spot around (X, Z) = (0 mm, 36 mm) corresponds to the target. The spatial resolution, which is evaluated from the size of the spot, is 0.20 mm (0.13 mm in the simulation) in the direction of X and 0.12 mm (0.11mm in the simulation) in the direction of Z.

Some artifacts can be observed in the image and the dynamic range of the image is 2.5dB lower than the 7 dB in the simulation. This suggests that the spatial arrangement of the transducers is not perfectly correct because of a manufacturing error.

This new concept of the 2-ch M-sequence encoding array, however, seems hopeful for realizing a simple and small size imaging system that will obtain an image in short time. In the near future the artifacts will be reduced by a certain correction of the geometric error in positioning the transducers or by applying a suitable operand.

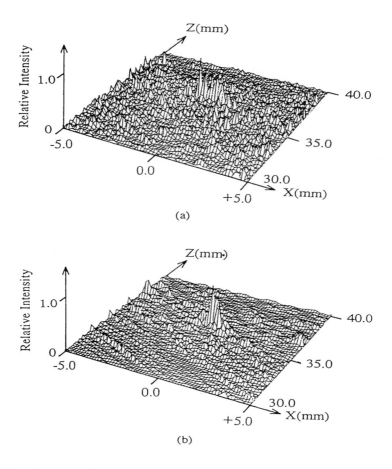

(a)

(b)

Figure 14 Two-dimensional point spread function for a single target with th 2-ch M-sequence encoding array transducer system: (a) pattern before filtering; (b) pattern after filtering.

A Multiple Shot 3-D Holographic Sonar Using a Set of Orthogonalized Modulating Signals

Data acquisition system

The most promising concept about the ultrasonic 3-D imaging system is that it is a real holographic sonar using a set of orthogonalized modulating signals. Incorporating a transducer array, the system is capable of high speed data acquisition by using transmitting pulses modulated with Walsh functions.

Walsh functions have a number of merits with respect to modulating the functions of acoustical imaging systems as follows:

- easy to generate and process because of their binary properties
- short duration time
- simultaneously obtains the echoes from targets in the observed 3-D space.

Figure 15 shows schematic drawings of the data acquisition system. Ultrasonic waves are simultaneously transmitted from n-transmitters. Sinusoidal waves of frequency f_o as carrier waves are modulated by binary codes obtained from the Walsh function synchronized to a clock signal. The period of the clock signals Δt is equal to an integral multiple of the sine wave's period $1/f_o$. The transmitting and receiving process is repeated n times. With the w-th transmitting and receiving cycle, the transmitting signal $u_p^{(w)}$ corresponds

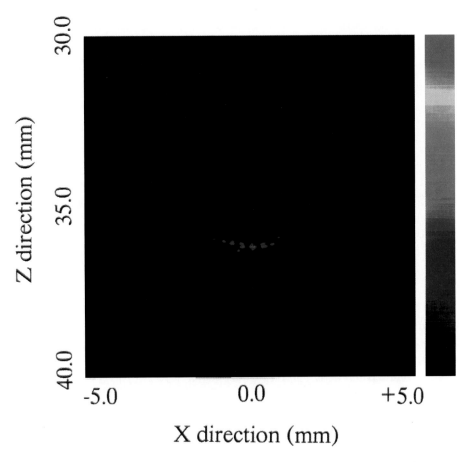

Photo 3 The cross-sectional image of a single target was reconstructed with the M-sequence encoding array transducer system following filtering. A correlation method was applied to the result. The imaging area was 10 mm × 10 mm centered on a point at X = 0 mm, Z = 35 mm.

to the p-th (p = 0, 1, ···, n-1) transmitter, and the waveform detected by i-th receiver is denoted by $u_p^{(w)}(t)$ and $r_i^{(w)}(t)$ respectively. The origins (t = 0) are set as at the start positions of each transmitting pulse for these functions.

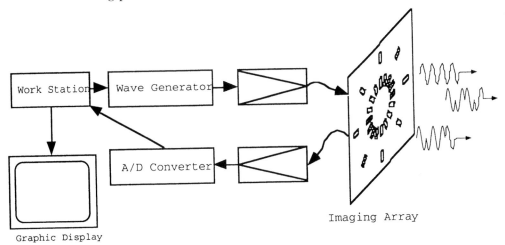

Figure 15 Holographic sonar system.

The transmitting waveform functions are given by the following expression:

$$u_p^{(w)}(t) = \sum_{i=1}^{n} W\zeta_{(p,w)i} f\big(t - (i-1)\Delta t\big) \text{ for } i = 1, 2, \cdots, n \tag{10}$$

where,

$$f(t) = \begin{cases} \exp(j2\pi f_o t) & 0 \leq t \leq \Delta t \\ 0 & \Delta t < t \end{cases} \tag{11}$$

is a sinusoidal pulse-burst of width Δt, W_{ij} denotes an i-j component of the $n \times n$ Hadamard matrix, and the column number $\zeta(p,w)$ has a one to one correspondence with the transmitter number p, that is:

$$\begin{cases} \zeta(p,w) = 1, 2, \cdots, n \\ \zeta(p,w) \neq \zeta(q,w) & \text{if } p \neq q \end{cases} \tag{12}$$

Reconstruction of image

The reconstructed object function S(x) corresponding to a position *x* is calculated from the sum of the correlations between the received echo functions and the expected signal for a single point target located at *x*. Therefore, S(x) is given by

$$S(x) = \left| \sum_{l=0}^{n-1} \sum_{i=0}^{m-1} \sum_{j=0}^{n-1} \phi_{ij}^{(l)} \big\{ \tau(\mu_i, x) + \tau(\upsilon_j, x) \big\} \right|^2 \tag{13}$$

where

$$\phi_{ij}^{(w)}(\tau) = \int r_i^{(w)}(t) u_i^{(w)}(t - \tau) * dt \tag{14}$$

is a cross correlation function between the transmitted signal $u_j^{(w)}$ at the j-th transmitter and received echo $r_i^{(w)}$ at the i-th receiver at the w-th cycle. The suffix "*" denotes a complex conjugate operator. $\tau(\gamma,x)$ is a delay (time of flight) between the target and the transducer at the position *x* and γ respectively.

Mod 2 repetition

Reducing speckled noises and spurious peaks in a reconstructed image due to errors, for example by arranging the transmitters and/or transducers, a method of transmission for multiple shots holography is proposed that is called "Mod 2 repetition". For comparison, another method called "Cyclic repetition" is also considered.

The column number function $\zeta(p,w)$ for Mod 2 repetition is given by:

$$\zeta(p,w) = p \;\&\; w + 1 \quad (p,w = 0, 1, 2, \cdots, \text{n-1}) \tag{15}$$

and for the "Cyclic repetition",

$$\zeta(p,w) = (p + w) \bmod(n) + 1 \quad (p,w = 0, 1, 2, \cdots, \text{n-1}) \tag{16}$$

where "&" expresses modulo 2 addition for every corresponding bit of represented binary numbers.

Orthogonal property of the transmitting pulses

The cross-correlation function for the i-th transmitting signal and j-th one is expressed by:

$$\phi_{ij}(\tau) = \sum_{l=0}^{n-1} \int u_i^{(l)}(t) u_j^{(l)}(t-\tau)^* \, dt = \begin{cases} n^2 \int f(t)f(t-\tau)^* \, dt & i = j \\ 0 & i \neq j \end{cases} \tag{17}$$

Transmitting signal for Mod 2 is

$$u_p^{(l)}(t) = \sum_{i=0}^{n-1} W_{(p\&l)i} f(1 - i\Delta t) \tag{18}$$

then, substituting $u_p^{(l)}(t)$ for $u_i^{(l)}(t)$ in the expression $\phi_{ij}(\tau)$ and rearranging it, the following expression can be obtained:

$$\phi_{ij}(\tau) = \sum_{l=0}^{n-1} \int \left\{ \sum_{i=0}^{n-1} W_{(p\&l)i} f(t - i\Delta t) \right\} \cdot \left\{ \sum_{j=0}^{n-1} W_{(p\&l)j} f(t - j\Delta t) \right\}^* \, dt \tag{19}$$

$$= \sum_{l=0}^{n-1} \sum_{i=0}^{n-1} \sum_{j=0}^{n-1} W_{pi} W_{li} W_{qj} W_{lj} \phi\{(j-i)\Delta t + \tau\}$$

Hadamard matrix's characteristic:

$$W_{(p\&l)i} = W_{pi} W_{li} \tag{20}$$

Here, applying other characteristics:

$$\sum_{l=0}^{n-1} W_{li} W_{lj} = n\delta_{ij} \tag{21}$$

Expression can be arranged as follows:

$$\phi_{ij}(\tau) = \sum_{i=0}^{n-1} \sum_{j=0}^{n-1} W_{pi} W_{qj} \left\{ \sum_{l=0}^{n-1} W_{li} W_{lj} \right\} \cdot \phi\{(j-i)\Delta t + \tau\} \tag{22}$$

$$= n \sum_{i=0}^{n-1} W_{pi} W_{qi} \delta_{ij} \phi(\tau) = n^2 \delta_{pq} \phi(\tau)$$

Where

$$\delta_{pq} \phi(\tau) = \begin{cases} n^2 \phi(\tau) & \text{for } p = q \\ 0 & p \neq q \end{cases} \tag{23}$$

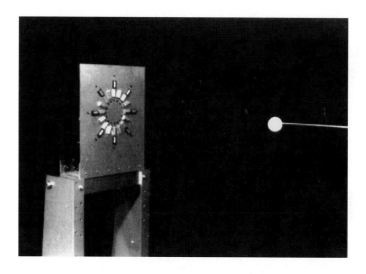

Photo 4 Appearance of the transducer array (and a sample target).

for $\tau < T_p$, where T_p is the period of repetition. Though the Mod 2 repetition is a simultaneous transmitting method, it is equivalent to a time-divided transmitting method in which an array transmits a single-pulse in turn from a single transmitter.

Experiments

Photo 4 shows the picture of the transducer array for the experiments. The array is arranged in dual circular forms of which the inner consists of 16 transmitters and the outer consists of 8 receivers. The frequency of carrier wave f_o is 50 kHz (wavelength $\lambda = 6.9$ mm and clock period Δt is $2 \times 1/f_o = 40$ µs).

Figure 16 shows a block diagram of the experimental system. Transmitting pulses generated by the wave-generator are reflected at objective targets and received by the array, then the signals are digitized by an A/D converter and transferred to an HP9000 workstation via a personal computer and the image is reconstructed.

Photo 5 shows the reconstructed images whose size is $64 \lambda \times 64 \lambda$ square. The intensities are increased 4 times so that the artifacts are more easily seen. The target is an acrylic rod of 15 mm in diameter set 1 m away from the array. Photo 5(a) shows the reconstructed image from only one transmitting and receiving process (single shot imaging) in which the artifacts are remarkable. Photo 5(c) shows the result following a Mod 2 repetition method.

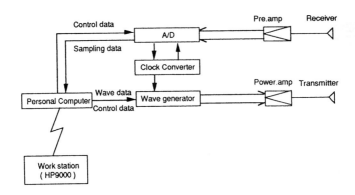

Figure 16 Block diagram of the experimental system.

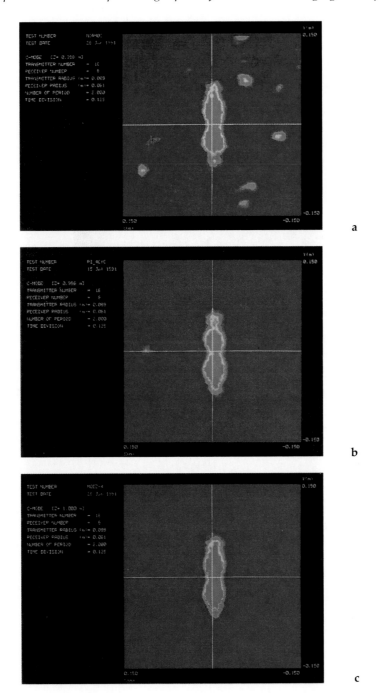

Photo 5 The reconstructed focal-plane images: (a) single shot transmitting; (b) simple cyclic transmitting; (c) Mod 2 transmitting.

Compared with the simple cyclic repetition (Photo 5(b)), the Mod 2 method has fewer artifacts.

Figure 17 shows the 3-D image by the Mod 2 method on a target made of acrylic plates formed into the Chinese characters "山大" which is the abbreviated form of "Yamagata University".

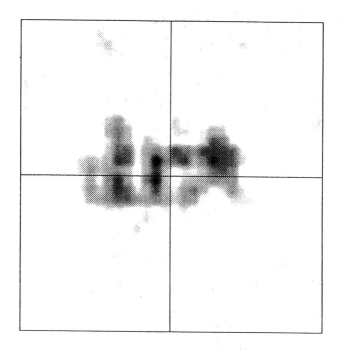

(a)

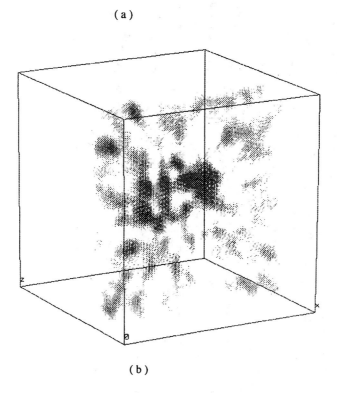

(b)

Figure 17 The reconstructed image for Mod 2: (a) focal-plane image; (b) 3-D image.

The ratios of the main peak to maximum subpeak are about 9 dB for the single shot imaging, and 16 dB for the Mod 2 method which is estimated from the simulation for 3-D boxel data for a single target set 1 m away from the array.

Thus, "the multiple-shot 3-D holographic sonar using a set of orthogonalized modulating signals" seems to be a very promising method for activating the development of real holographic imaging systems as short and medium distance sonars for underwater imaging.

Theoretically and experimentally refining this method, the resolution and range of underwater imaging will progress and improve remarkably.

Conclusion

In this review the discussion has concentrated on the imaging systems used to obtain acoustic images in the vertical or focal plane similar to a photograph, but adding the lateral which is typically presented by conventional sonars.

The types of acoustic imaging systems using transducer array basically are classified into two categories:

- beamformed or beam scanning systems
- holographic approach — reconstructing the hologram

The first method "beamformed" is outlined as two types of conventional sonars to express the distinction from the latter method "holographic". The "beamformed" methods are discussed as existing synthetic aperture array systems based on the classifications:

- Synthetic aperture side scan sonar
- Multibeam scanning sonar

The principle of acoustic imaging techniques on newly developed holographic sonars were reviewed as follows:

- Encoding aperture using an inclined linear array
- 2-ch M-sequence encoding array transducer system
- Multiple shot 3-D holographic sonar using a set of orthogonalized modulating signals

Among the newly developed methods, "multiple shot 3-D holographic sonar using a set of orthogonalized modulating signals" seems to be a very promising method for activating the development of real holographic imaging systems as short and medium distance sonars for underwater imaging. The resolution and range of underwater imaging will improve and progress remarkably by refining the above method.

References

Nitadori, K., Mano, K., and Kamata, H.: An experimental underwater acoustic imaging system using multi-beam scanning, *Acoustical Imaging,* Vol. 8, 1980.

Sutton, J. L.: Underwater acoustic imaging, *Proc. IEEE,* Vol. 67, No. 4, April 1979.

Tamura, Y.: An echolocation using spatial modulation and wide-band pulse, *Proc. SICE,* Vol. 26, 1987.

Tamura, Y. and Akatsuka, T.: An echolocation and imaging using transducers of directionally distinguishable impulse response, *Acoustical Imaging,* Vol. 18, 1991.

chapter twelve

Visualization of Sound Radiation and Propagation by the Sound Intensity Method

Hideki Tachibana, Hiroo Yano and Yoshito Hidaka

Institute of Industrial Science, University of Tokyo
Roppongi 7-22-1, Minato-ku, Tokyo, 106 Japan

Abstract — The sound intensity (SI) method is a new measurement technique by which sound intensity, the sound power passing through unit area, can be directly measured as a vector quantity. In this paper, the principle of this method is briefly explained and typical examples of its application to various acoustic measurements regarding sound radiation and propagation are introduced by focusing on sound field visualization.

Introduction

Sound pressure and particle velocity are the most essential quantities prescribing a sound field; they correspond to voltage and electric current respectively, in an electric system. As electric power is the product of voltage and electric current, the sound intensity is the product of sound pressure and particle velocity and it means the acoustic power passing through a unit area in a sound field.

Although the definition of sound intensity is very simple as mentioned above, the method of measuring this quantity has not been realized for a long time because it was very difficult to measure the particle velocity simultaneously with the sound pressure. Owing to the recent development of such technologies as transducer production and digital signal processing, it has finally been realized (Fahy 1989).

According to the sound intensity (SI) method, the sound power flow in an arbitrary sound field can be directly measured as a vector quantity. In this paper, the principle of the SI method is briefly explained and some examples of its applications are introduced by focusing on sound field visualization.

Principles of Measurement of Sound Intensity

Sound intensity is defined as,

$$\vec{I} = \overline{p(t)\vec{u}(t)} \qquad (1)$$

where $p(t)$ is sound pressure, $\overrightarrow{u(t)}$ is particle velocity and $^{-}$ denotes time-averaging.

In the sound field consisting only of a plane wave, there exists the following relationship between $p(t)$ and $u(t)$:

$$u(t) = p(t)/\rho c \tag{2}$$

where ρ is the density of air and c is the sound velocity. In this case, the absolute value of the sound intensity can be obtained only from the sound pressure as follows:

$$I = \left|\vec{I}\right| = \overline{p^2(t)}/\rho c \tag{3}$$

However, Equation (2) is not necessarily valid for general sound fields and sound intensity must be obtained according to Equation (1). For this point, the measurement of particle velocity is a serious problem. It is very difficult to measure the particle velocity with the correct phase relationship with the sound pressure and this has prevented the realization of the sound intensity measurement for a long time.

Regarding this problem, T. Schultz invented an epoch-making idea called "2-microphone method" in 1956 (Fahy 1989). In this method, the second term of Equation (4) is approximated by the finite difference between the sound pressures at two points closely spaced as expressed by Equation (5) and consequently the particle velocity is approximated as expressed by Equation (6):

$$\rho \frac{\partial u}{\partial t} + \frac{\partial p}{\partial r} = 0 \tag{4}$$

$$\frac{\partial p}{\partial r} \doteqdot \frac{p_2(t) - p_1(t)}{\Delta r} \tag{5}$$

$$u_r(t) \doteqdot -\frac{1}{\rho \Delta r} \int_{-\infty}^{t} \{p_2(t) - p_1(t)\} dt \tag{6}$$

where $p_1(t)$ and $p_2(t)$ are the sound pressures at the two points and Δr is the separation distance between them. Consequently, the sound intensity component in the direction of r can be approximated as follows:

$$I_r \doteqdot -\frac{1}{\rho \Delta r} \overline{\frac{p_1(t) + p_2(t)}{2} \int_{-\infty}^{t} \{p_2(t) - p_1(t)\} dt} \tag{7}$$

Figure 1(a) shows the actual measurement system based on the principle mentioned above. This method is called the "direct method". In this case, the two channels must have the identical performances in both the amplitude and phase characteristics for all of the components such as the pressure microphones, amplifiers, and band pass filters. This point has been the most serious problem for sound intensity measurement. Fortunately, it has almost been solved by recent developments in transducer manufacturing and digital signal processing technologies.

By expressing Equation (7) in the frequency domain, we have:

$$I_r(f_1 - f_2) \doteqdot -\frac{1}{2\pi\rho\Delta r} \int_{f_1}^{f_2} \frac{\text{Im}\{G_{12}(f)\}}{f} df \tag{8}$$

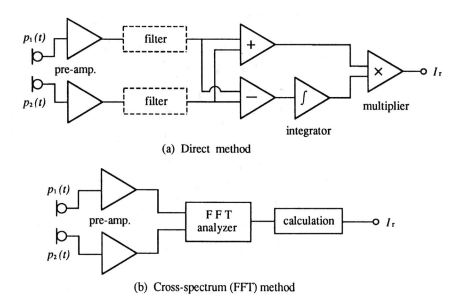

Figure 1 Sound intensity measurement systems by 2-microphone method: (a) Direct method; (b) Cross-spectrum (FFT) method.

where $\mathrm{Im}\{G_{12}(f)\}$ is the imaginary part of the cross-spectrum density function between $p_1(t)$ and $p_2(t)$. The calculation according to this formula can be easily realized by use of FFT analyzer as shown in Figure 1(b). This method is called "cross-spectrum method" or "FFT method".

As a sound intensity probe, two pressure-type condenser microphones with identical amplitude and phase characteristics are used in side-by-side or face-to-face configurations as shown in Figure 2. Figure 3 shows an example of a 3-D intensity probe consisting of three sets of 2-microphone probes, by which the x, y, and z components of the sound intensity vector can be measured at the same time.

Applications of Sound Intensity Method

Sound Radiation Measurements

One of the most significant applications of the SI method is the measurement of the sound radiation characteristics of sound sources. Some examples of this kind of application are presented below:

1. **Sound radiation from a vibrating plate** (Hidaka et al. 1987) — As a basic study, the sound intensity radiation from a vibrating strip of a vinyl chloride plate was measured. As shown in Figure 4, it was excited at the center point and the sound intensity in the near field and the vibration velocity on the plate were measured. From these results we can see that the "source" positions from which the sound is radiated and the "sink" positions into which the sound is absorbed are arrayed alternately.

Figure 2 2-microphone intensity probes: (a) side by side; (b) face to face.

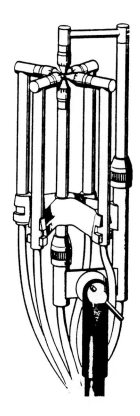

Figure 3   3-D intensity probe (B&K).

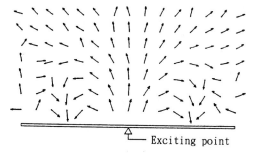

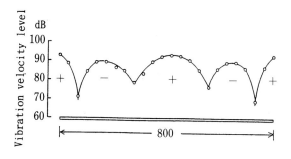

Figure 4   Sound radiation from a vibrating plate.

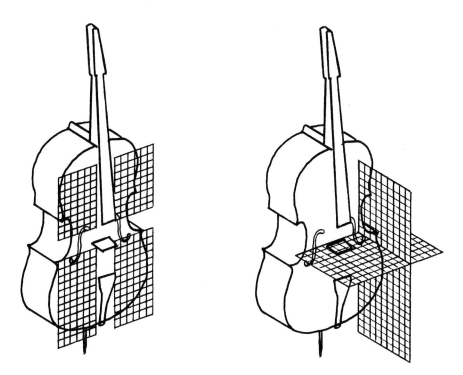

Figure 5 Measurement of sound power flow from a violoncello: (a) parallel to the sound board; (b) perpendicular to the sound board.

2. **Sound radiation from a violoncello** (Tachibana and Hidaka 1991) — Musical instruments are very interesting subjects for acoustics and many investigations have been made to date. We have measured the sound power radiation characteristics of various musical instruments by the SI method. Among them, the sound intensity radiation patterns from a violoncello are shown here. In this measurement, the second string (d, open) was played by a bow which was automatically moved by using a reversible motor and the sound intensity distributions were obtained in each 1/3 octave band as shown in Figure 5. As a result, Figure 6 shows the contour maps of sound intensity normal to a plane parallel to the sound board. In the case of 160 Hz band in which the fundamental tone of 147 Hz is included, the sound radiation pattern is uniform. On the other hand, in the cases of 315 Hz and 630 Hz bands in which the second and fourth harmonic tones are included, not only the positive sound intensity (solid lines) but also the negative sound intensity (dotted lines) are seen and the radiation patterns are much more complicated. Figure 7 shows the sound intensity vector maps measured in two sections perpendicular to the sound board through the f-hole. Among these results, it should be noted that sound power is absorbed in the lower part of the body in the case of 315 Hz band.

3. **Sound radiation from building structures** (Yano et al. 1990) — Building structures can radiate sound when they vibrate. As a model experiment investigating this kind of problem, an I-shaped steel beam 120 mm × 60 mm and 1 m long was hung in an anechoic room and the sound intensity around it was measured by exciting an end of the beam. In this experiment, the measurement was conducted both with and without a vibration damping treatment on the web of the beam. In Figure 8(a) and (b), the solid lines indicate the sound intensity vectors measured by stationary random excitation and the dotted lines indicate those measured by impulsive

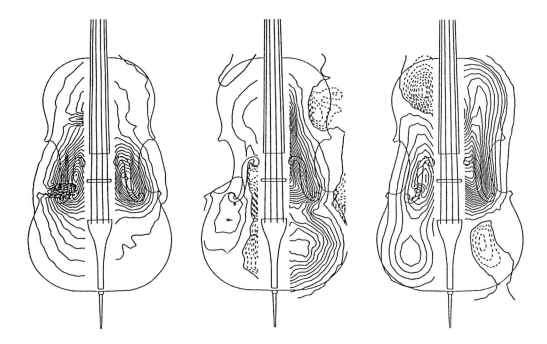

Figure 6 Normal sound intensity contour on the plane parallel to the sound board (1/3 octave band): (a) 160 Hz band; (b) 315 Hz band; (c) 630 Hz band.

excitation. Each vector was obtained from the sound intensities measured in x and y directions. Almost all results measured by the two excitation methods are in good agreement with and without damping treatment. From these results, the radiated sound power per unit length of the beam was calculated by integrating the sound intensity component normal to the measurement line. Figure 9 shows the efficiency of the vibration damping treatment, which was obtained as the difference of the sound power with and without damping treatment.

4. **Sound radiation from rotating automobile tires** (Oshino and Tachibana 1991) — In order to investigate the sound radiation characteristics of rolling automobile tires, a field experiment was performed by applying the SI method. In this experiment, a trailer equipped with a probe scanning machine was driven on a test course and the normal intensity distribution on the plane of 1 m × 1 m area, which was set parallel to the rear tire, was measured in actual running conditions. Figure 10(a) shows the measured results for a rib tire, where we can see that the front and rear parts of the contact patch are the dominant sound sources. Figure 10(b) shows the measured results for a lug tire, where we can see that the contact patch is the main sound source and the sound radiation pattern is much more complicated than in the case of the rib tire.

Sound Field Analysis

1. **Sound diffraction around a cylinder** (Hidaka et al. 1987) — As a very simple example of the measurement of sound diffraction made by the SI method, Figure 11 shows the sound intensity flow around a rigid cylinder. For 500 Hz as shown in (a), the wavelength of the sound is larger than the diameter of the cylinder (34 cm) and it is clearly seen that the sound is diffracted behind the cylinder. On the other hand,

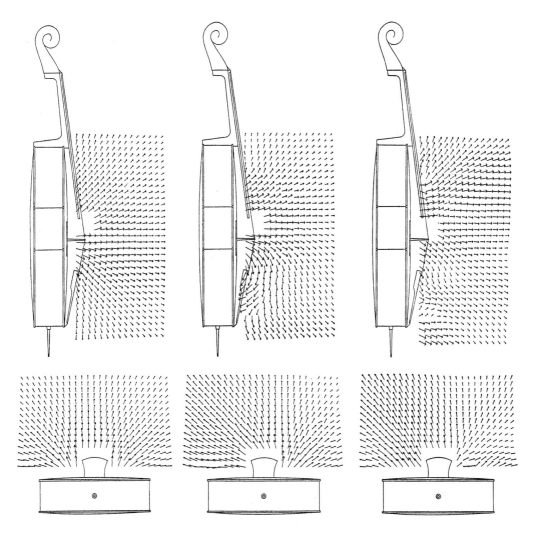

Figure 7 Sound intensity vector map on the planes perpendicular to the sound board (1/3 octave band): (a) 160 Hz band; (b) 315 Hz band; (c) 1250 Hz band.

for 2 kHz as shown in (b), the wavelength is smaller than the diameter of the cylinder and the sound intensity behind the cylinder is very weak in comparison.

2. **Sound diffraction over an office screen** (Hidaka et al. 1993) — Low partition walls are often used in offices to provide visual and acoustical privacy. Figure 12 shows the sound intensity flow over a low partition wall in a room measured by the SI method. This experiment was performed using a 1/2 scale model.

3. **The effect of sound diffusers** (Tachibana 1988) — In auditoriums, walls are often made shaped irregularly to make the sound field diffusive. As an experimental study to investigate the effect of such acoustic treatments, the sound reflection from a plane with a triangular projection and that with a semicircular projection were measured by the SI method. Figure 13(a) shows the sound intensity flow of 500 Hz over the two reflecting planes. In this case, the wavelength is larger than the scale of the section of the projections and the sound intensity reflection patterns are almost the same. In both cases, sound intensity vortices are seen. On the other hand, Figure 13(b) shows the sound intensity flow of 8 kHz. In this case, the wavelength is smaller than the scale of the projections and the two reflection patterns are much

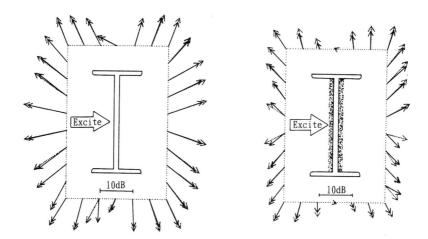

Figure 8 Sound radiation from I-shaped steel beam (4 kHz in 1/3 octave band): (a) without damping treatment; (b) with damping treatment.

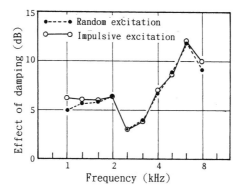

Figure 9 The effect of the damping treatment.

different. These results indicate that the effect of sound diffusion is different in high frequencies between the two types of projections.

4. **Measurement of acoustic diffusiveness in rooms** (Hidaka et al.) — As a trial to observe the extent of acoustic diffusiveness in a room, the 3-D sound intensity measurement using an impulsive sound source was applied in various rooms. Among the results, Figure 14 shows the vector loci of instantaneous sound intensity on the horizontal (x-y) plane measured in a concert hall (a) and in a reverberation room (b). In the former case, the direct sound from the sound source (x-direction) and discrete early reflections from right and left directions are clearly seen. In the latter case, the reflections after the direct sound are omnidirectional; it indicates that the sound field is much diffusive.

Sound Insulation Measurement (Yano and Tachibana 1987)

The SI method can be effectively applied to sound insulation measurement. As an example, Figure 15 shows the measured results of sound power flow transmitting through a window. In this measurement, the sound source was located inside the room and the sound

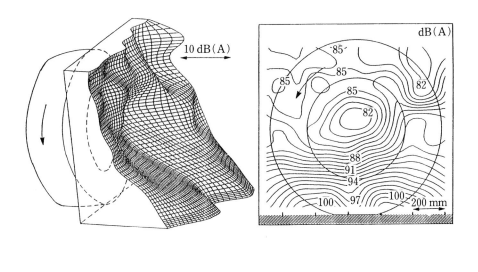

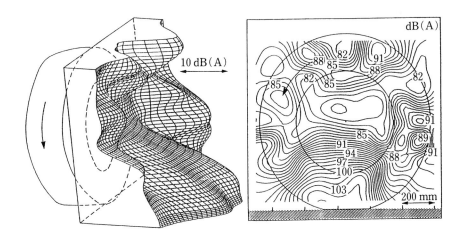

Figure 10   Sound intensity radiation from rolling tires (A-weighted): (a) rib tire; (b) lug tire.

intensity normal to the window was measured outside at many discrete points. From these results, we can see that sound power uniformly transmits through the window in the case of low frequency (250 Hz), as shown in (a), whereas sound power transmits dominantly through the edge parts of the window in the case of high frequency (2 kHz), as shown in (b). As is clearly seen in this example, the SI method is very effective for the detection of acoustic weak points in sound insulation of building walls.

Conclusions

In this paper, some examples of acoustic radiation, propagation, and transmission measurements made by the SI method have been introduced by focusing on sound field visualization. According to the SI method, sound pressure and particle velocity can be measured simultaneously and almost all acoustic quantities can be obtained. Therefore, this acoustic measurement method will be much more widely applied in the future.

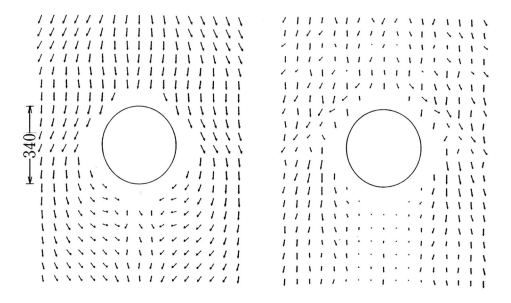

Figure 11   Sound diffraction around a cylinder: (a) 500 Hz; (b) 2 kHz.

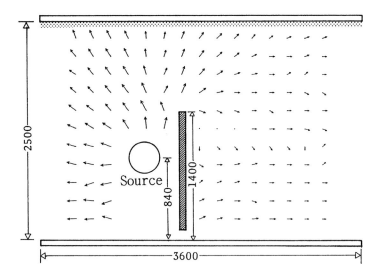

Figure 12   Sound diffraction around a low partition (2 kHz in 1/2 scale model experiment).

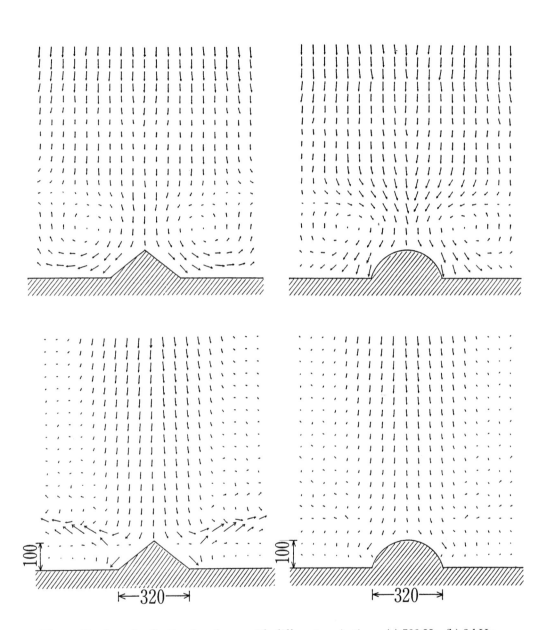

Figure 13 Sound reflection by planes with different projections: (a) 500 Hz; (b) 8 kHz.

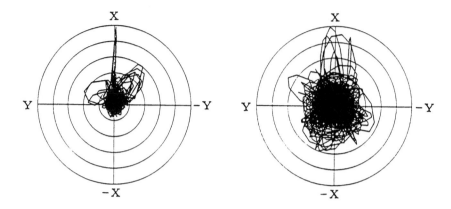

Figure 14 Vector loci of instantaneous sound intensity on horizontal plane: (a) concert hall; (b) reverberation room.

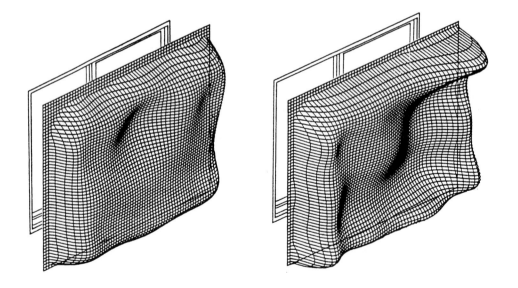

Figure 15 Sound intensity transmitting through a window: (a) 250 Hz in octave band; (b) 2 kHz in octave band.

References

Fahy, F. J., 1989: *Sound Intensity*, Elsevier Applied Publishers.

Hidaka, Y., Ankyu, S., and Tachibana, H., 1987: Sound field analyses by complex sound intensity, *J. Acoust. Soc. Jpn. (J)*, Vol. 43 No. 12, pp. 994–1000.

Hidaka, Y., Yano, H., and Tachibana, H., 1993: Analysis of room acoustics by introducing 3D sound intensity measurement, *Proc. Acoust. Soc. Jpn.* Autumn Meeting of A.S.J., pp. 793–794.

Oshino, Y. and Tachibana, H., 1991: Noise source identification of rolling tires by sound intensity measurement, *J. Acoust. Soc. Jpn. (E)*, Vol. 12 No. 2, pp. 87–92.

Tachibana, H., 1988: Visualization of sound fields by sound intensity technique, *Proc. 2nd Symp. Acoust. Intensity*, pp. 117–126.

Tachibana, H. and Hidaka, Y., 1991: Visualization of sound radiation from a violoncello, *J. Acoust. Soc. Jpn. (J)*, Vol. 46 No. 10, pp. 864–866.

Yano, H., Hidaka, Y., and Tachibana, H., 1990: Sound and vibration measurements by impulsive excitation, *J. Acoust. Soc. Jpn. (E)*, Vol. 11 No. 2, pp. 77–82.

Yano, H. and Tachibana, H., 1987: Applications of sound intensity technique to architectural acoustic measurement, *J. Acoust. Soc. Jpn. (J)*, Vol. 43 No. 12, pp. 966–974.

Index